ABOUT THE BOOK

This book contains no formal proofs in regards to shapes, formulas and theorems used in the solutions to problems contained. The reader will find that some problems may be more challenging than typical geometry problems. This is not a textbook for geometry, but rather a workbook for individuals who want to test their knowledge of the subject.

There are 268 problems that include standard 2-dimensional shapes such as circles, rectangles, triangles, ellipses, trapezoids, parallelograms, quadrilaterals, rhombuses, pentagons, and more. Complex 3-dimensional shapes such as prisms, pyramids, cuboids, cones, ellipsoids, spheres, and cylinders are also used among other interesting forms. Also, some shapes may be enclosed in other shapes to add flavor to the subject. Most figures will require one to find the area, volume, line segments and angles with respect to given information. Techniques of Trigonometry, Pythagorean Theorem, direct proportion, and logic are heavily implemented in order to solve any given problem. In addition, application problems (real-life problems) will require one to use the standard Cartesian plane and geometric shapes to solve them.

The book opens with typical questions regarding the lines, planes and surfaces. Some problems may be given as word problems requiring one to draw pictures or analyze the given information. Most of the problems include accompanying figures, because geometry by definition works best through visual illustrations when any given information is used. Answers are shown in a multiple choice format. Some answers may be similar due to certain units that are used in the process of calculating areas and volume.

Detailed solutions immediately follow the problems. Some solutions will include figures with added content (units) or figure parts for better comprehension of the problem. Units used will be kilometers, centimeters, meters, feet, miles, yards, inches, and more. Final answers will often be required to be converted from one unit to another, including square and cubic units. Some problems will also require one to use algebra before arriving at the answer, and some will include questions that enforce knowledge of coordinate geometry (2-dimensional or 3-dimensional).

ABOUT THE AUTHOR

David Alaverdian started his life journey as a musician. He grew passion for mathematics from an early age, and geometry was one of his major obsessions. After earning his undergraduate and graduate music degrees from The Juilliard School and Manhattan School of Music in NYC, he completed his second undergraduate degree (Bachelor of Science) in Applied Mathematics. He became enthusiastic about writing math books ever since. He is currently pursuing an actuarial career, and also offers one-on-one tutoring services during his spare time. His rich tutoring experience reaches beyond 15 years. He has helped countless number of students improve their grade in K-12 and Calculus math courses.

HOW TO USE THE BOOK

For best results, try to work out each problem before looking at their respective solutions. If a figure is missing in the problem, try to draw a picture. Most complex problems will include a figure to assist the reader. If stuck in the problem, think about where you might be confused and why. Do not immediately look at the solutions. It may be some formulas that hinder the completion of any solution step. To help better understand major concepts, we include basic formulas and unit conversions below.

FORMULAS AND UNIT CONVERSIONS

1. For any right triangle, the **Pythagorean Theorem** states the relationships of its sides as

$c^2 = a^2 + b^2$, where c is a hypotenuse (longest side and opposite side of the triangle's 90-degree angle).

2. For any triangle the **Law of Sines** states that

$$\frac{a}{\sin(A)} = \frac{b}{\sin(B)} = \frac{c}{\sin(C)}$$

where

a is the opposite side of angle A,

b is the opposite side of angle B

c is the opposite side of angle C.

3. For any triangle the **Law of Cosines** states that

$c^2 = a^2 + b^2 - 2ab(\cos(C))$, where c is an opposite side to angle C.

$a^2 = b^2 + c^2 - 2bc(\cos(A))$, where a is an opposite side to angle A.

$b^2 = a^2 + c^2 - 2ac(\cos(B))$, where b is an opposite side to angle B.

4. Some common units are given below.

1 m = 100 cm = 1,000 mm = 0.001 km

1 m^2 = 10,000 cm^2 = 1,000,000 mm^2 = 0.000001 km^2

1 m^3 = 1,000,000 cm^3 = 1,000,000,000 mm^3 = 0.000000001 km^3

1 m = 3.280839895 ft = 1.093613298 yd

1 m^2 = 10.76391042 ft^2 = 1.195990046 yd^2

1 m^3 = 35.31466672 ft^3 = 1.307950618 yd^3

1 in = 0.083333333 ft = 2.54 cm

1 in^2 = 0.006944444 ft^2 = 6.4516 cm^2

1 in^3 = 0.000578704 ft^3 = 16.387064 cm^3

1 km = 1,000 m

1 km^2 = 1,000,000 m^2

1 km^3 = 1,000,000,000 m^3

1 ft = 12 in = 0.3048 m = 30.48 cm

1 ft^2 = 144 in^2 = 0.09290304 m^2 = 929.0304 cm^2

1 ft^3 = 1,728 in^3 = 0.028316847 m^3 = 28,316.84659 cm^3

1 mL = 1 cm^3

1 gal = 3.78541 L = 3,785.41 mL = 4 qt = 8 pt

1 mi = 5,280 ft

1 mi^2 = 27,878,400 ft^2

π radians = 180°

5. Formulas for shapes are given below.

Circle circumference: $2\pi r$, where r is the radius

Circle area: πr^2

Triangle area: $\frac{bh}{2}$, where b is the base side and h is the height

Equilateral triangle area: $\frac{\sqrt{3}s^2}{4}$, where s is the side of the triangle

Square area: s^2, where s is the square side. The area is also the square of its diagonal divided by 2 (or the product of its diagonals divided by 2)

Square perimeter: $4s$

Rectangle perimeter: $2l + 2w$, where l is the length and w is the width

Rectangle area: $l \cdot w$

Rhombus area: $\frac{sh}{2}$, where s is the side and h is the height. The area is also the product of its diagonals divided by 2 (this also works for the area of a kite).

General Quadrilateral area: $\sqrt{(s-a)(s-b)(s-c)(s-d)-abcd(\cos^2(\frac{A+B}{2}))}$ where *a*,*b*,*c*,*d* are quadrilateral sides, $s = \frac{a+b+c+d}{2}$, and <A and <B are opposite interior angles

Right trapezoid area: $\frac{h(a+b)}{2}$, where *a* is one base, *b* is the second base and *h* is the height

Ellipse area: $\pi(ab)$, where *a* and *b* are mutually perpendicular principal radiuses

Regular pentagon area: $a \bullet \frac{\sqrt{5(5+2\sqrt{5})}}{4}$, where *a* is the side

Regular hexagon area: $\frac{3a^2\sqrt{3}}{2}$, where *a* is the side

Sphere circumference: $4\pi r^2$, where *r* is the radius

Sphere volume: $\frac{4}{3}\pi r^3$

Ellipsoid surface area: $4\pi\left(\frac{(ab)^{1.6}+(ac)^{1.6}+(bc)^{1.6}}{3}\right)^{\frac{1}{1.6}}$ where *a*, *b* and *c* are principal radiuses

Ellipsoid volume: $\frac{4\pi}{3} \bullet abc$, where *a*, *b* and *c* are principal radiuses

Right circular cone surface area: $\pi r(r+\sqrt{h^2+r^2})$ where *r* is the radius of the base and *h* is the vertical height

Right circular cone volume: $\frac{\pi r^2 h}{3}$, where *r* is the radius

Pyramid volume (right and oblique): $\frac{lwh}{3}$, where *l* is the length, *w* is the width and *h* is the vertical height from the base to the apex of the pyramid

Right pyramid surface area: $lw + w\sqrt{(\frac{l}{2})^2 + h^2} + l\sqrt{(\frac{w}{2})^2 + h^2}$, where l is length, w is width and h is the vertical height joining the apex and the centroid of the pyramid base

Closed right cylinder surface area: $2\pi r^2 + 2\pi rh$, where r is the radius and h is the height

Cylinder volume (right and oblique): $\pi r^2 h$, where h is the vertical height. For oblique cylinder, $h = h_s(\sin(a))$ where h_s is its slant height, and a is the angle between its slant height and circular base

Closed oblique cylinder surface area: $2\pi r^2 + 2\pi rh_s$, where h_s is the slant height

Torus surface area: $(2\pi r)(2\pi R)$, where r is the outer radius and R is the inner radius

Torus volume: $(\pi r^2)(2\pi R)$, where r is the outer radius and R is the inner radius

6. Trigonometric identities for the right triangle are listed below.

$$\sin x = \frac{\text{opposite side of angle } x}{\text{hypotenuse}}$$

$$\cos x = \frac{\text{adjacent side of angle } x}{\text{hypotenuse}}$$

$$\tan x = \frac{\text{opposite side of angle } x}{\text{adjacent side of angle } x}$$

$$\csc x = \frac{\text{hypotenuse}}{\text{opposite side of angle } x} = \frac{1}{\sin(x)}$$

$$\sec x = \frac{\text{hypotenuse}}{\text{adjacent side of angle } x} = \frac{1}{\cos(x)}$$

$$\cot x = \frac{\text{adjacent side of angle } x}{\text{opposite side of angle } x} = \frac{1}{\tan(x)}$$

For the following right triangle, angle x can be expressed using inverse trigonometric functions, namely arccosine ($\cos^{-1}$), arcsine ($\sin^{-1}$) or arctangent ($\tan^{-1}$):

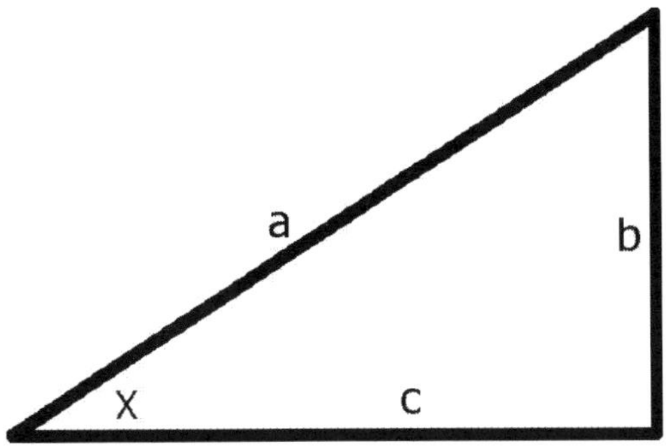

$m\angle x = \sin^{-1}\left(\frac{b}{a}\right) = \cos^{-1}\left(\frac{c}{a}\right) = \tan^{-1}\left(\frac{b}{c}\right)$

7. Trigonometric angle addition formulas are listed below.

$\sin(a + b) = \sin(a)\cos(b) + \sin(b)\cos(a)$

$\sin(a - b) = \sin(a)\cos(b) - \sin(b)\cos(a)$

$\cos(a + b) = \cos(a)\cos(b) - \sin(a)\sin(b)$

$\cos(a - b) = \cos(a)\cos(b) + \sin(a)\sin(b)$

$$\tan(a + b) = \frac{\tan(a) + \tan(b)}{1 - \tan(a)\tan(b)}$$

$$\tan(a - b) = \frac{\tan(a) - \tan(b)}{1 + \tan(a)\tan(b)}$$

8. For any polygon with the number of sides $s \geq 3$, the interior angle sum is **180(*s* – 2)**. If the polygon is equilateral, the number of degrees for each of the *s* interior angles is $\frac{180(s-2)}{s}$.

In a circle, an inscribed angle is half the degree of the arc spanned by this angle.

The central angle has the same degree measure as its spanned arc.

Two chords intersecting in a circle form an angle that measures one-half the sum of its spanned arcs.

Two secant lines, one secant line and one tangent line, or two tangent lines forming an angle outside the circle make this angle measure one-half the difference of its spanned arcs.

A point outside the circle from which two tangent lines are drawn makes their lengths equivalent.

9. The slope-intercept form equation of a straight line is $\boldsymbol{y = mx + b}$, where m is the slope of the line, and b is its y-intercept. The slope of the line is $\boldsymbol{m} = \frac{y_1 - y_0}{x_1 - x_0}$, where (x_o,y_o) and (x_1,y_1) are the two known points on the line determining its slope. The point-slope form of the line is $\boldsymbol{y = m(x - x_o) + y_o}$, where (x_o,y_o) is the known point on the line. Distance from a line $\boldsymbol{ax + by + c = 0}$ to the point (x_o,y_o) is $\frac{|ax_o + by_o + c|}{\sqrt{a^2 + b^2}}$. This distance is a line segment *perpendicular* to the line $ax + by + c = 0$.

10. Arithmetic sequence is a set of increasing or decreasing terms that are a constant distance apart from one another. For example, a sequence 1,3,5,7,9 is an arithmetic sequence because the integers are (3 – 1) = (5 – 3) = (7 – 5) = (9 – 7) = 2 integers apart (constant distance) from one another. If the first term of such sequence is a_1, the nth term of the sequence is $\boldsymbol{a_n = a_1 + (n - 1)d}$, where a_1 is the first term, n is the total number of terms, and d is the constant distance between any two adjacent terms. The **sum** of the first n terms in an arithmetic sequence is $\boldsymbol{S} = \frac{n(a_1 + a_n)}{2} = \frac{n(n+1)}{2}$.

11. A slope of the line or curve defined by function f(x) at $x = a$ is $\lim_{h\to 0}\frac{f(a+h) - f(a)}{h}$. For example, the slope of the line f(x) = -12x + 4 can be found by simply locating -12 (the slope) using slope-intercept form, or using the formula mentioned above as $\lim_{h\to 0}\frac{-12(a+h) + 4 - (-12a + 4)}{h} = \lim_{h\to 0}\frac{-12h}{h}$ = -12. When it comes to curves, there are infinitely many slopes (because a curve is not a straight line). For this reason, the limit formula (also called derivative formula) is especially helpful.

NOTE TO THE READER

You will find many problems contained in this book useful for your geometry class. While some problems may be different from typical class assignments one may encounter, most questions address the standard theorems found in geometry. This book may also be a great resource to have in your Kindle collection.

I apologize ahead for any errors you may find. Please send an e-mail to **liszt6@hotmail.com** if you feel there is an error anywhere in the text. However, only legitimate cases will be considered for review.

Hope you enjoy this book, and best of luck in your geometry studies!

David Alaverdian

---

268 SOLVED PROBLEMS

1. Of the 5 angles, 3 are supplementary and the remaining 2 are complementary. What is the degree measure sum of all the 5 angles?

A. 90

B. 135

C. 270

D. 225

Essentially, it does not matter how many angles are supplementary or complementary together. It could have been 10 angles that are complementary and 1 million angles that are supplementary. In the end, it simply matters what complementary and supplementary means. Complementary means the angles make a 90 degree angle together, while supplementary means that angles make 180 degrees together. So, putting it together we have 90 + 180 = 270 (a 270 degree angle measure sum).

Answer: C

2. A plane can be determined in all possible cases by

A. one point

B. two non-intersecting straight lines

C. three non-collinear points

D. any two points

We can eliminate choices A and D, because one point cannot determine a plane, and any two points can lie on many planes. Imagine a line between any two points and a third line perpendicular to that line. You can rotate the perpendicular line by 360 degrees about the 2-point line, which means any two points can have more than one plane. Now, choice B may seem closer to the truth, however, even though any parallel straight lines do not intersect (and may definitely form a plane, like an X-Y plane), in a three-dimensional space the non-intersecting straight lines may not form a plane (if they are not parallel). The only choice is to find 3 non-collinear points (not all three points on the same line), which definitely forms a plane in all possible cases.

Answer: C

3. Two intersecting planes form a

A. plane

B. point

C. straight line

D. two-dimensional surface

Intersecting planes cannot form another plane or one single point. There are infinite points created by intersecting planes. These points form a straight line. The easiest way to see this is to imagine a cube. Any two adjacent faces of the cube are the two distinct planes, and they meet at their shared boundary, thus producing a line. This is valid for more complex prisms as well. Thus, choices A, B and D are eliminated (choice D is a plane).

Answer: C

4. Two intersecting straight lines share

A. no points

B. infinite number of points

C. a plane

D. a straight line

When straight lines intersect, they will definitely share one point. They cannot share infinite number of points, because they can intersect only once. The same can be said about sharing a line (line can have an infinite number of points). They do, however, share a plane, because intersecting straight lines satisfy the 3 non-collinear point definitions.

Answer: C

5. Line *C* is perpendicular to a plane when it is

A. perpendicular to one line on the plane

B. perpendicular to all lines on the plane containing the intersection point with *C*

C. parallel to the plane

D. intersecting the non-collinear points on the plane

A line in question can be perpendicular to a line on the plane in infinite ways. For example, it can lie on the plane and still be perpendicular to the line the plane. This is demonstrated in question 2. A line cannot intersect non-collinear points, since it can only intersect the plane at one point. A line cannot be parallel to the plane, since it will no longer be perpendicular to that plane. Now, if one designates a point of intersection on the plane for the line, and all the lines on the plane share this point of intersection, then the line in question will definitely be perpendicular to the plane.

Answer: B

6. What has a zero dimension?

A. line segment

B. circle

C. plane

D. point

A line segment has only one dimension, whereas circles and planes are two-dimensional.

Answer: D

7. An exterior angle in a triangle must be

A. the sum of the two non-adjacent interior angles

B. supplementary with the adjacent interior angle

C. all of the above

D. none of the above

Exterior angles must always make 180 degrees with an adjacent interior angle in a triangle. They also equal the sum of the two non-adjacent angles in a triangle.

Answer: C

8. A right angle cannot be formed by

A. two complementary angles

B. two acute angles

C. two intersecting lines

D. two supplementary angles

A right angle measures 90 degrees. Two acute angles can form it. Acute means less than 90 degrees. Complementary angles always make a 90-degree angle. Two intersecting lines can form a 90-degree angle if they are perpendicular to each other. Two supplementary angles (which make a 180-degree angle) cannot form it.

Answer: D

9. Vertical angles are all but

A. acute

B. obtuse

C. adjacent

D. supplementary

Vertical angles are mirrors of each other, so they can be acute, obtuse and supplementary. This simply depends on whether the intersecting lines are perpendicular to each other. However, vertical angles can never be adjacent.

Answer: C

10. Two points form a plane if

A. a line is perpendicular to one of them

B. a third point that does not lie on the line between them is added

C. they are collinear

D. a third point that lies on the line between them is added

Two points alone cannot form a plane. If one adds a third point not lying on the line between the original two points, then the plane is formed.

Answer: B

11. In the following figure, lines *m* and *n* are parallel, and line *p* is a transversal. What must be the degree measure of angle *C*?

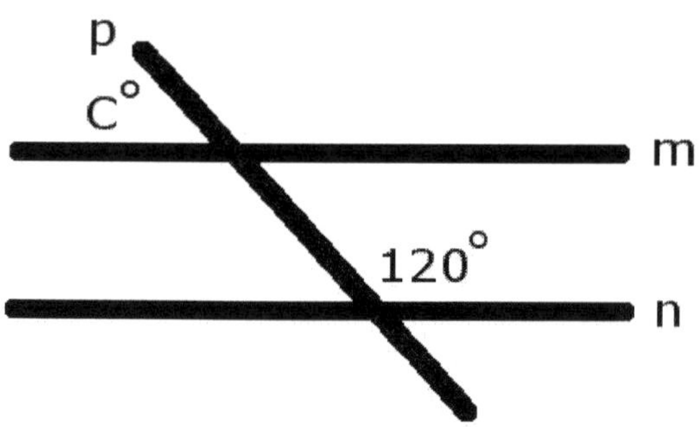

A. 65 degrees

B. 60 degrees

C. 75 degrees

D. 55 degrees

Since the supplement of 120 degrees is 60 degrees (for the interior angle adjacent to the 120 degree angle), and angle C is identical to this supplement, it must also be 60 degrees.

Answer: B

12. In the following figure, lines *c* and *d* are parallel, and line *m* is a transversal. Which angles are equal?

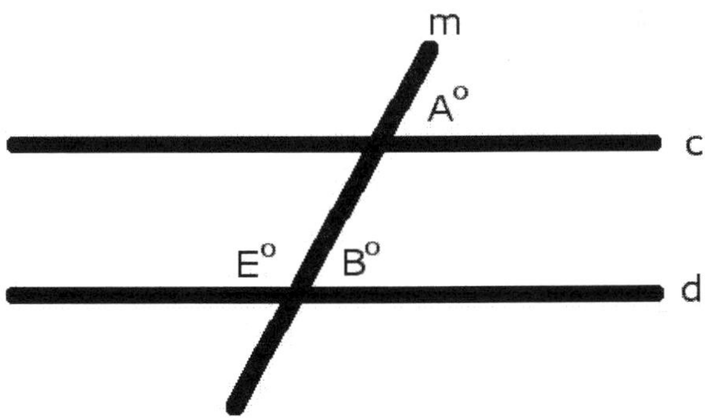

A. angles *E* and *A*

B. angles *B* and *E*

C. all of the above

D. none of the above

Measure of angle A = measure of angle B, so choices A, B, and C are eliminated.

Answer: D

13. In the figure below, lines *a* and *b* are perpendicular to each other, and line *c* bisects angle *adb*. What is the measure of angle *X*?

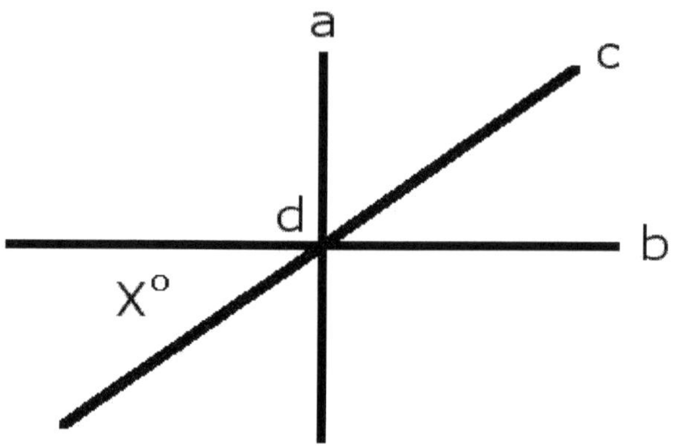

A. 90 degrees

B. 75 degrees

C. 65 degrees

D. 45 degrees

Since lines *a* and *b* are perpendicular to each other, measure of angle *adb* equals 90 degrees. Line *c* bisects this angle, making angle *cdb* measure 45 degrees. This angle is a vertical angle (mirror image) of angle X.

Answer: D

14. In the following figure, lines *M* and *N* are perpendicular to each other. Ray *C* bisects angle *NDM*. What is the degree measure of angle *CDA*?

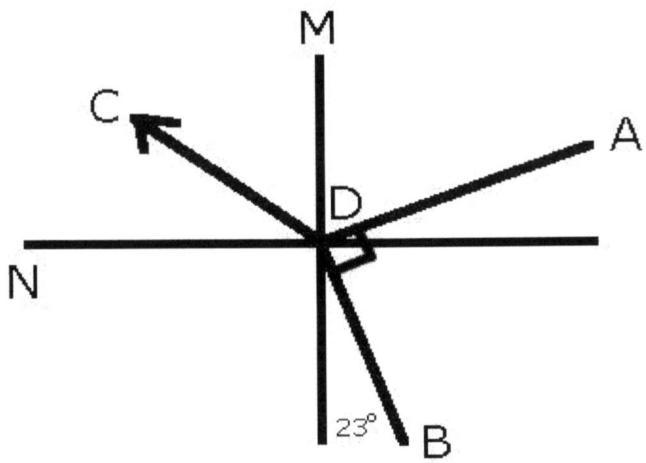

A. 135 degrees

B. 112 degrees

C. 113 degrees

D. 120 degrees

Since angle ADB measures 90 degrees, measure of angle MDA = 180 – (23 + 90) = 180 – 113 = 67 degrees. Since ray C bisects angle NDM, and lines M and N are perpendicular to each other, measure of angle CDM = 45 degrees. Thus, measure of angle CDA = 67 + 45 = 112 degrees.

Answer: B

15. If the lines intersect at the center in the rectangle below, what is the sum of measures of angles *X* and *Y*?

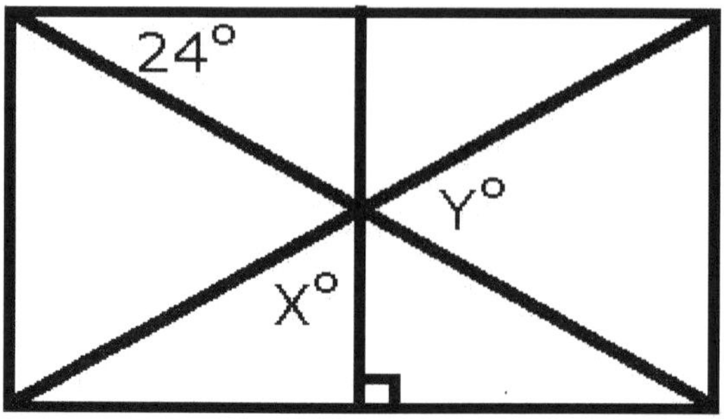

A. 114 degrees

B. 112 degrees

C. 100 degrees

D. 116 degrees

The simplest way to solve this problem is to find the adjacent angle for both *X* and *Y*, and then subtract it from 180 to get *X* + *Y* measure. To do this, first notice that the center vertical line makes a right angle with the rectangle length. This makes two angles in the top left triangle known. Now we must find the third angle in that triangle, and this angle will be the vertical angle for the adjacent angle of *X* and *Y*. This angle measures 180 – (24 + 90) = 180 – 114 = 66 degrees. Thus, *X* + *Y* measure 180 – 66 = 114 degrees.

Answer: A

16. In the following figure, a blue circle is inscribed in a square whose side m is 5 cm long. The area of the white non-shaded square region is 2π square centimeters. What is the circumference of the circle (to the nearest hundredth)?

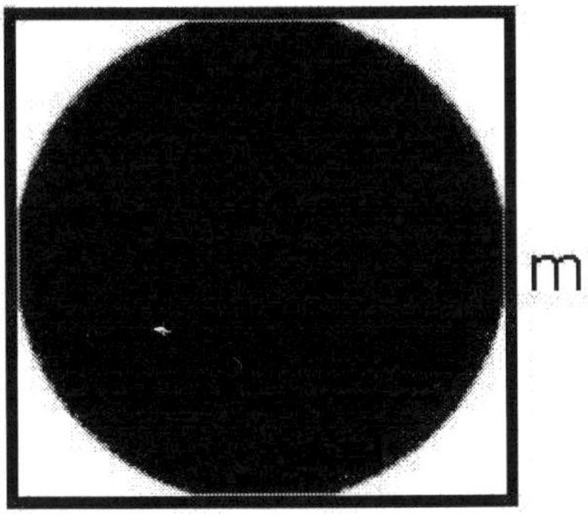

A. 15.34 cm

B. 16.02 cm

C. 8.09 cm

D. 12.58 cm

Area of the square is $5 \cdot 5 = 25$ cm ^{2}. This makes the area of the circle to be $(25 - 2\pi)$ cm ^{2}. Now, we set an equation for the area of the circle: $\pi r^2 = (25 - 2\pi)$, so that $r = 2.440849679$ (always use the complete answer given in the calculator for the intermediate steps). Thus, the circumference of the circle is $2\pi r = 2\pi(2.440849679) \approx 15.34$ cm.

Answer: A

17. The following figure shows an equilateral triangle, with side 8 ft, being inscribed in a circle with center point *A*, whose area is 25π square feet. What is the area of the region bounded by *ABCDA* (to the nearest hundredth)?

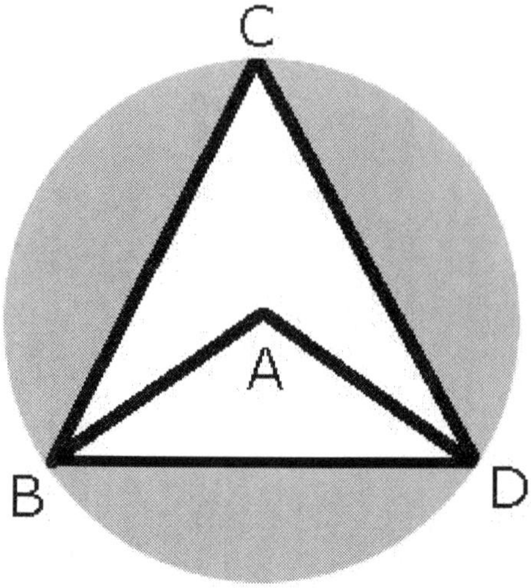

A. 14.04 ft^2

B. 12.78 ft^2

C. 9.86 ft^2

D. 15.71 ft^2

To get the area of the region bounded by ABCDA, we must subtract the area of the region bounded by ABDA from the area of the equilateral triangle. First, the area of the triangle is $\frac{8^2\sqrt{3}}{4}$ = 27.71281292. Since the region bounded by ABDA is an isosceles triangle (2 sides equal to the radius of the circle), the 3rd side BD can be divided in half, and this way we can find the altitude to A from the segment BD. Since the half of BD is 4, and AB is 5, the altitude is $\sqrt{5^2 - 4^2}$ = 3 feet. We used the famous *Pythagorean Theorem* to find the altitude.

There was also a way to avoid this theorem, by simply noticing that this is a 3-4-5 triangle. Now, the area of the triangle ABDA is $\frac{bh}{2} = \frac{8 \bullet 3}{2}$ = 12 square feet. Thus, the area of ABCDA is 27.71281292 – 12 ≈ 15.71 ft^2.

Answer: D

18. In the following figure, two circles are inscribed in a rectangle whose length is 20 meters. Find the area of the top half of the rectangle region not included in the circles (to the nearest hundredth).

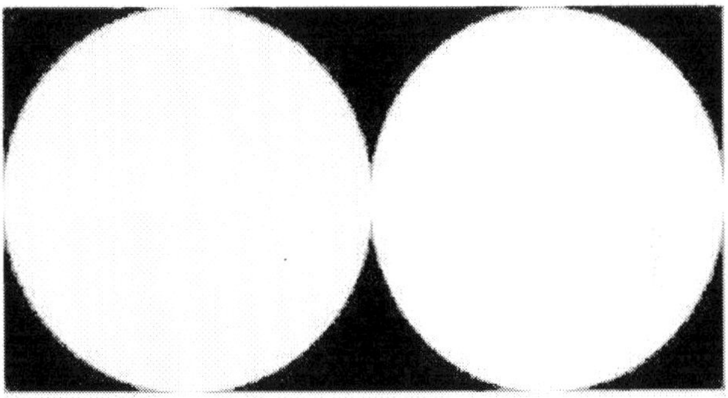

A. 42.92 square meters

B. 51.07 square meters

C. 45.23 square meters

D. 21.46 square meters

The rectangle length of 20 meters equals the 2 diameters of the enclosed circles, which means that the radius of each circle is 5 meters. Also, this gives the width of the rectangle which is 10 meters long. We are asked to find the area of the rectangle region not included in the two circles. To do this, we must subtract the area of the two circles from the area of the rectangle. The area of the 2 circles is $2(5^2)\pi = 157.0796327$. The area of the rectangle is $20 \cdot 10 = 200$. Thus, the desired area of the top portion of the rectangle region not bounded by the circles is $\frac{200 - 157.0796327}{2} \approx 21.46$ square meters.

Answer: D

19. The following figure shows 2 adjacent identical equilateral triangles, with side 10 feet, making a rhombus. What is the area of the rhombus formed (to the nearest hundredth)?

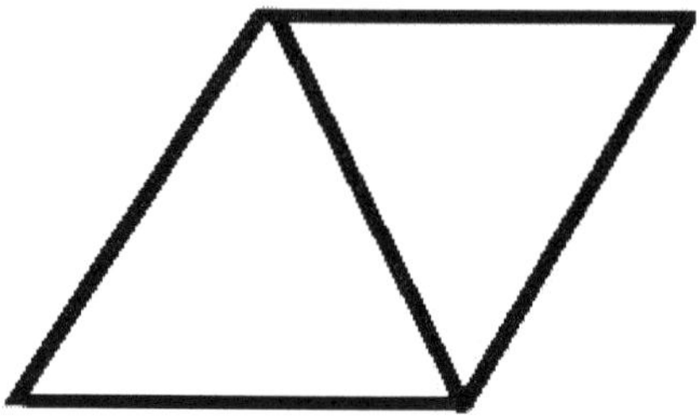

A. 86.60 ft^2

B. 89.01 ft^2

C. 84.59 ft^2

D. 82.67 ft^2

The easiest way is to find the area of one of the equilateral triangles and then simply multiply it by two. The area is $\frac{10^2\sqrt{3}}{4}$ = 43.30127019. Thus the area of the rhombus is 86.60 ft^2 (rounded to the nearest hundredth).

The other longer way would be to find the altitude of one of the triangles and then use it in the area formula for the rhombus (formula we use for the area of a rectangle). We use the half of the side of the triangle, and then use the Pythagorean Theorem to find the altitude. The altitude is $\sqrt{10^2 - 5^2} = \sqrt{75}$ or 8.660254038. Thus, the area of the rhombus is (*base* • *height*) = 10(8.660254038) ≈ 86.60 ft^2.

Answer: A

20. The following figure shows a trapezoid enclosing an isosceles triangle whose base is 16 feet long. If the height of the trapezoid is 8 feet long, and its area is 144 square feet, find the perimeter of the white non-shaded trapezoid region.

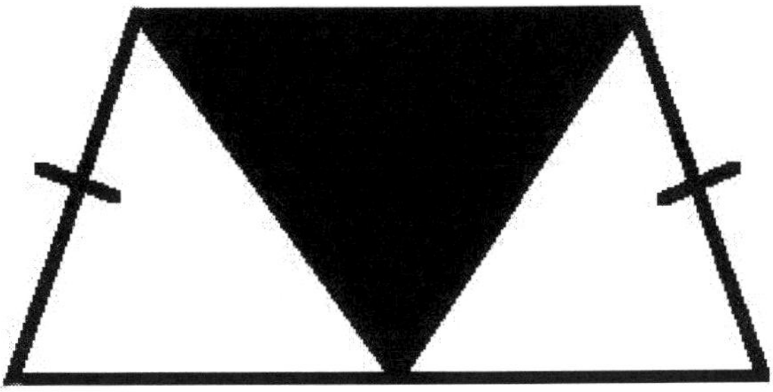

A. 54 feet

B. 56 feet

C. 58 feet

D. 59 feet

This problem involves quite a number of steps one needs to take in order to arrive at the answer. The perimeter we need to find does not include the base of the triangle enclosed in the trapezoid. So, we must 'remove' it from the trapezoid while finding the triangle's 2 equal sides other than its base (the triangle is isosceles).

To find one of the 2 equal sides, we need to use the *Pythagorean Theorem*. We know the height of the triangle (which is also the height of the trapezoid), so we divide the triangle base side in half ($\frac{16}{2}$) and use the height to get one of the 2 equal sides. We have the third side to be $\sqrt{8^2 + 8^2}$ = 11.3137085. Now, we must find the longest side of the trapezoid (its base). We know the formula for the area of the trapezoid: $\frac{(base1 + base2) \bullet height}{2}$. So, we simply solve for the *base2* (bottom side). We have 144 = $\frac{(16 + base2) \bullet 8}{2}$, and we get *base2* = 20. We are now only missing the 2 congruent sides of the trapezoid for perimeter. We can divide the trapezoid base side in half ($\frac{20}{2}$) because the enclosed triangle is isosceles, and since the top horizontal side half (8) of the trapezoid is shorter than the base side (10), we subtract the shortest side from the longest: 10 – 8 = 2. This distance is the additional distance that makes the trapezoid bottom base longer than the top base and crucial to find the two congruent trapezoid sides (while also using the known height). Again, we use the *P. T.*, so that the required slanted side is $\sqrt{2^2 + 8^2}$ = 8.246211251. We now have all the required sides for our perimeter. Thus, the perimeter is 2(8.246211251) + 2(11.3137085) + 20 ≈ 59 feet.

Answer: D

21. The following figure shows an open right cylinder (on both ends) whose diameter is 10 feet long, and height 17 feet long. What is the outer surface area of this cylinder?

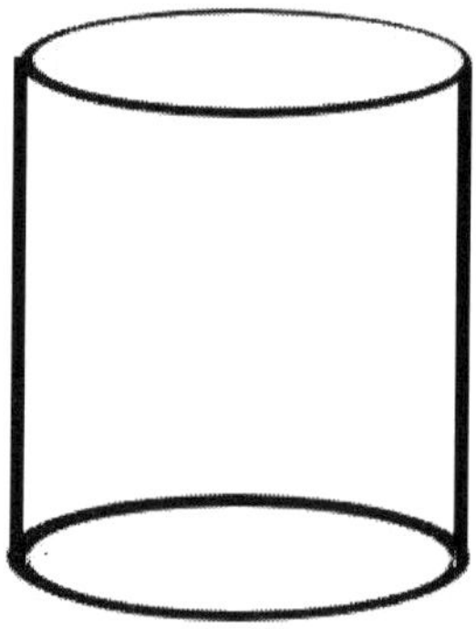

A. 534 ft^2

B. 269 ft^2

C. 538 ft^2

D. 267 ft^2

We are asked to find the surface area of the cylinder that does not include the areas of the two circles that close it on both sides, so that it is simply $2\pi rh = 2\pi(5)(17) \approx 534$ square feet.

Answer: A

22. The following figure shows a rhombus being enclosed into a rectangle whose length is 9 feet and width 6 feet. What is the area (in inches) of the rectangle region that does not include the rhombus?

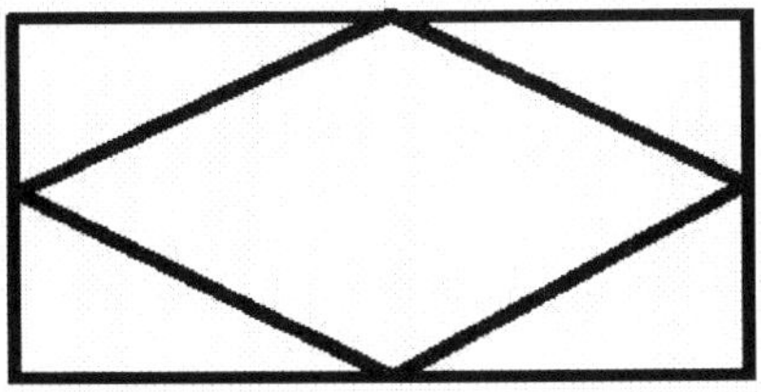

A. 3888 square inches

B. 7776 square inches

C. 54 square inches

D. 27 square inches

If one knows the diagonals of the rhombus, the area is known. In this case, the two diagonals equal the length and height of the rectangle that inscribes the rhombus. So, the area is $\frac{diag1 \bullet diag2}{2} = \frac{6 \bullet 9}{2} = 27$. Now, the area of the rectangle is (6 • 9) = 54. So, the area of the rectangle region not included in the rhombus is 54 – 27 = 27 square feet. Now, we must be careful to convert the answer to inches. 1 square foot = 144 square inches (12 • 12), so our answer is 144 • 27 = 3,888 square inches.

Answer: A

23. The following figure shows a part of a circle attached to a square whose side is 10 yards. Find the area and the perimeter of the figure.

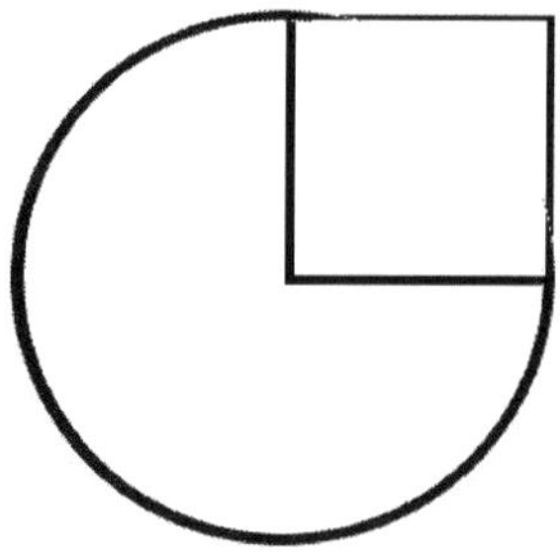

A. A = 75π + 100, P = 15π + 20

B. A = 15π + 20, P = 75π + 100

C. A = 75π + 20, P = 15π + 100

D. A = 15π + 100, P = 75π + 20

We must find the value of three-quarters of the circle's circumference and add it with the half of the perimeter of the square. The square side is the radius of the circle, which is 10. So, $\frac{3}{4}$ of the circle circumference is $(\frac{3}{4})(2)(\pi)(10) = 15\pi$. Adding it with the half of the square perimeter gives 15π + 20. The area of the figure is done similarly, except this time we will need the entire square area added to the $\frac{3}{4}$ of the circle area. The circle area is $(\frac{3}{4})(100\pi) = 75\pi$. Adding it to the area of the square gives 75π + 100.

Answer: A

24. The following figure shows a small equilateral triangle inside a large equilateral triangle. The ratio of the areas of the triangles is 3:1. If the area of the larger triangle is 81, find the side of the smaller triangle.

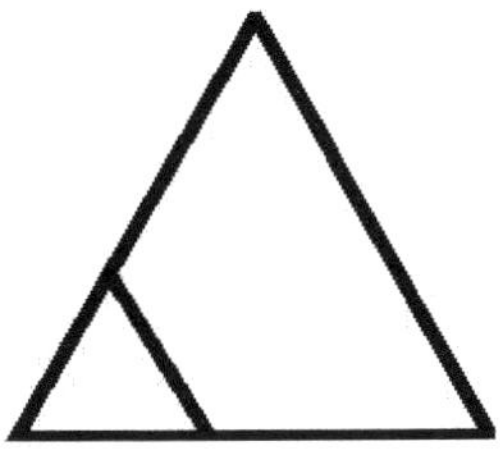

A. 6

B. 7

C. 8

D. 9

Since the ratio of the areas of the large triangle to small triangle is 3:1, we have 81 = 3*A*, which means that *A* (area of the small triangle) is 27. Now we use the area formula for equilateral triangle to find the side: $27 = \frac{s^2\sqrt{3}}{4}$, which gives $s \approx 8$.

Answer: C

25. The following figure shows one isosceles triangle inside another isosceles triangle. The two congruent base parts are each 6 meters in length. If the area of the larger triangle is 144 square meters, and its height 14 meters, find leg *x*.

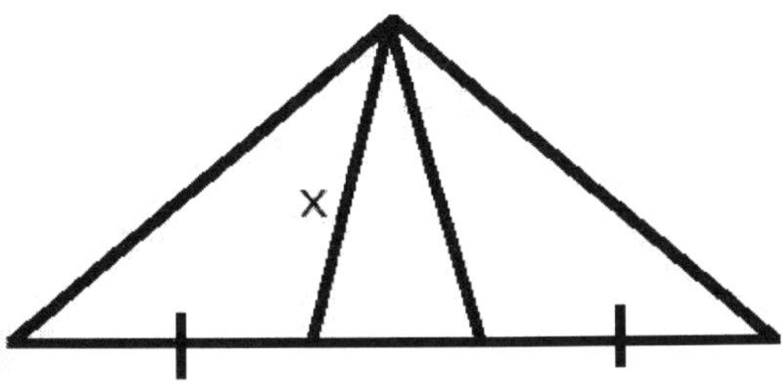

A. 13 m

B. 11 m

C. 15 m

D. 16 m

We are missing a part of the base of the large isosceles triangle, so we will name it *y*. The area of the large isosceles triangle is 144 = $\frac{bh}{2}$, and our base will be 6 + 6 + *y* = 12 + *y*. The area is then 144 = $\frac{(12 + y)(14)}{2}$, which gives *y* = 8.571428571. Now, since the smaller triangle is also isosceles, we can divide our newly found *y* in half to get 4.285714286. Now we use the *Pythagorean Theorem* to get leg *x*, so that our *x* = $\sqrt{4.285714286^2 + 14^2}$ ≈ 15 m.

Answer: C

26. The following figure shows a cube with sides painted with identical circles whose radius is 5 feet. What is the surface area (in yards) of the cube regions where no circles are painted?

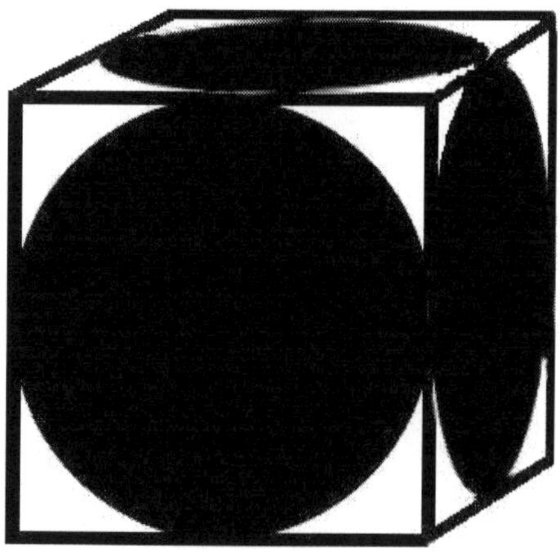

A. 14 square yards

B. 129 square yards

C. 450 square yards

D. 150 square yards

We must find the surface area of the cube while excluding the painted circles on all six faces of the cube whose edges are tangent to these circles. Since the radius of each circle is 5, each side of the square is 10. The surface area of the cube is 6(10)(10) = 600. The surface area of the six circles is 6(25π) = 150π. Thus, the desired surface area of the cube not including the areas of the circles is (600 – 150π) square feet.

Now we must be careful to convert our answer to yards. Since 1 square yard = 9 square feet, our desired area is $\left(\frac{\sqrt{600-150\pi}}{3}\right)^2$ ≈ 14 square yards. The easiest method to convert the units backwards is to simply divide (600 – 150π) by 9 to get 14.30678911.

Answer: A

27. In the following figure, a sliced arc of the circle is 1.5π. What is the measure of central angle *Y*?

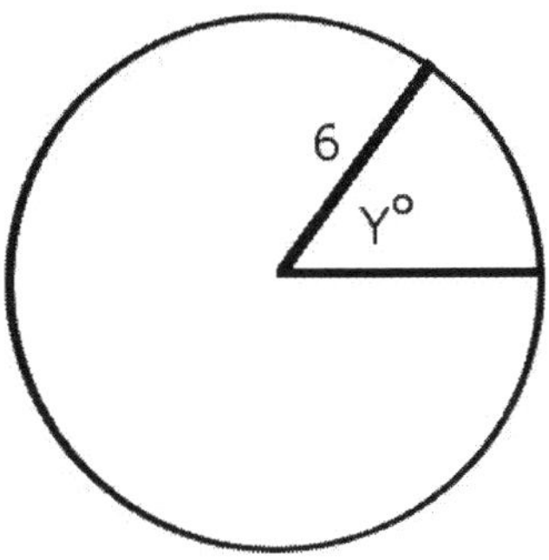

A. 35 degrees

B. 60 degrees

C. 55 degrees

D. 45 degrees

In this problem, we use the proportion of the sliced arc to the whole circumference to find the measure of angle *Y* (which encompasses the sliced arc) with respect to 360 degrees. The circumference of the circle is $2(\pi r) = 2(6)(\pi) = 12\pi$.

Setting up the direct proportion, we get $\frac{1.5\pi}{12\pi} = \frac{1}{8}$, so that our angle *Y* measure is $\frac{1}{8}$ of 360 degrees, namely 45 degrees.

Answer: D

28. In the following figure, the center of the circle joins the right and isosceles triangles. The area of the isosceles triangle is 44. The area of the circle is 121π. What is the value of expression (*x* + *y*)?

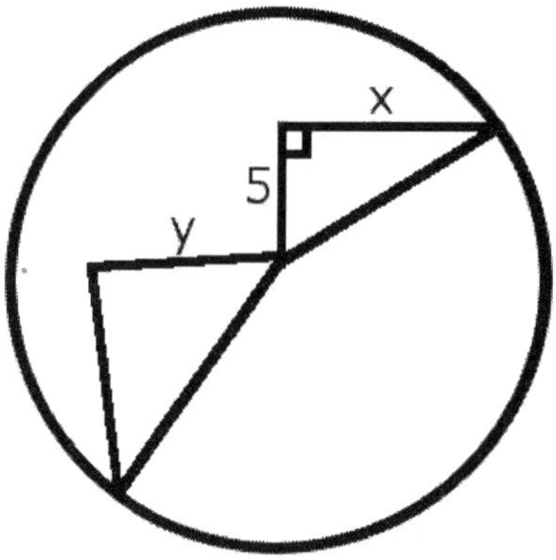

A. 18

B. 17

C. 22

D. 20

The hypotenuse of the right triangle and base of the isosceles triangle both equal to the radius of the circle. Since the area of the circle is 121π, the radius must be 11. So, first we find *x* using the *Pythagorean Theorem*: $x = \sqrt{11^2 - 5^2} = 9.797958971$.

Now, the area of the isosceles triangle is 44, so we use the area formula: $44 = \frac{11h}{2}$, solving for the height, and $h = 8$. To get y, we divide the base in half $(\frac{11}{2})$ and use the *Pythagorean Theorem*: $y = \sqrt{5.5^2 + 8^2} = 9.708243919$. Thus, $x + y \approx 20$.

Answer: D

29. In the following figure, the two lines *a* and *b* are parallel. Which of the following equations is valid?

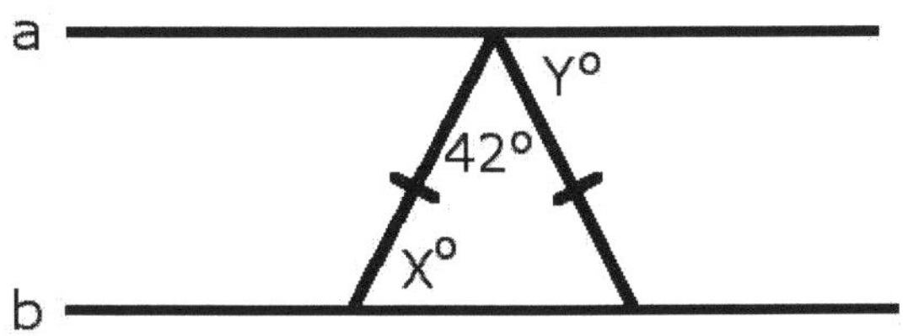

A. $({}_{m<}x - 180) = -({}_{m<}y + 42)$

B. $({}_{m<}x + 42) = 138$

C. $42 = ({}_{m<}x - {}_{m<}y)$

D. $2_{m<}x = (180 - {}_{m<}y)$

To check individual choice and its corresponding expressions, we must simplify them first. Re-writing the first expression (choice A), we get $x + y = 180 - 42 = 138$.

Now, since angle *Y* measure equals the unknown third angle measure (this angle is also equal to *X* in measure, why?) inside the triangle (because *a1* and *a2* lines are parallel), the sum of angles *x*, *y* and 42 must equal in measure to 180 degrees. We see that it is indeed true, since $x + y + 42 = 138 + 42 = 180$. No need to check the other answer choices, but you may do so for your own practice.

Answer: A

30. In the following graph, if the slope of the red line segment is 2.5, what is the value of the hypotenuse (to the nearest 0.1)?

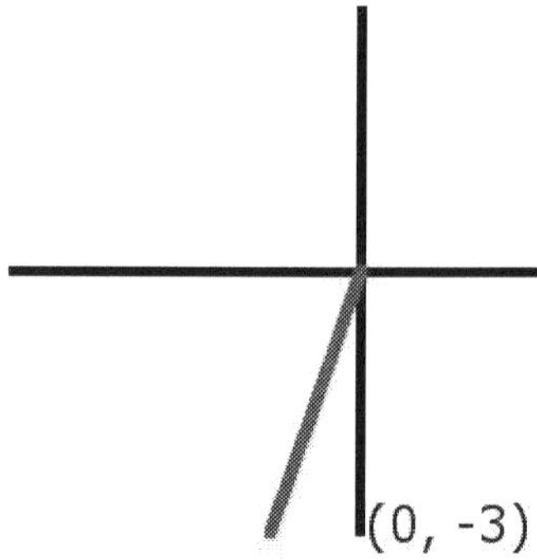

A. 3.4

B. 3.0

C. 2.9

D. 3.2

The slope means: change of *y* divided by the change of *x*. To find the hypotenuse (the red line segment), we must find the *x*-coordinate (the missing side). We already know the *y*-coordinate, which is 3 in absolute value.

To find the missing side (we can label it *S*), we use the property of the slope mentioned above. We have 2.5 = $\frac{3}{S}$, so that *S* = 1.2. Now we use the *Pythagorean Theorem* to find the hypotenuse *h*: *h* = $\sqrt{1.2^2 + 3^2} \approx 3.2$.

Answer: D

31. The following figure shows an isosceles triangle whose side is tangent to the circle. The circle radius is equal to the one of the identical sides of the triangle. If the circumference of the circle is 31π feet, find the measure (in inches) of the height drawn from the hypotenuse to the vertex of the isosceles triangle.

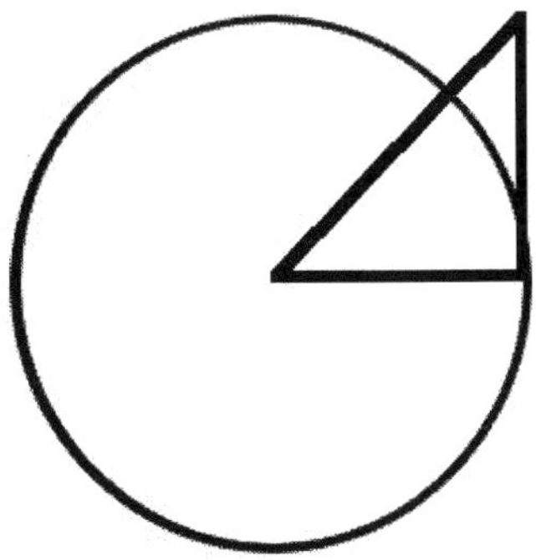

A. 132 in

B. 96 in

C. 79 in

D. 124 in

The circumference of the circle is 31π, which means that the radius is 15.5. Since both identical sides are 15.5 (the radius is equal to one of these sides), the hypotenuse is $\sqrt{2\,(15.5^2)}$ = 21.92031022. Since the triangle is isosceles, we can divide it in half, making the hypotenuse (or base of the triangle) $\frac{21.92031022}{2}$ = 10.96015511. We now have a right triangle created with two dimensions 15.5 (the radius is now a hypotenuse of this right triangle) and 10.96015511. The height of this triangle is the height we need to find. This is simply $\sqrt{15.5^2 - 10.96015511^2}$ = 10.96015511 feet. You probably see that it is equal in value to the half of the hypotenuse of the isosceles triangle. This is because the height that we needed to find cuts the vertex angle (90-degree angle) in half, making 45 degrees. Since the angle at the circle center is also 45 degrees (why?), the sides opposite of these two angles must be equal. This explains the result. In inches the answer is 10.96015511(12) ≈ 132.

Answer: A

32. The area of the rectangle shown is 346. The following statements are thrown:

I. The product of the diagonals (not shown) of the rectangle is 346.

II. The product of diagonals (not shown) of the enclosed rhombus is 346.

III. The area of the rhombus is half the area of the rectangle

IV. The longest diagonal of the rhombus equals the diagonal of the rectangle

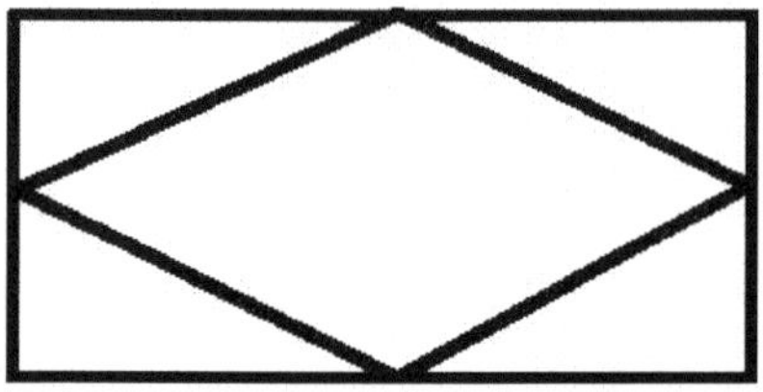

Which of the statements above is NOT true?

A. I & II only

B. I and III only

C. I and IV only

D. I, II and IV

Area of the rectangle is 346, which is known by multiplying the width times the length. Statement I is false, because the diagonals of the rectangle will exceed the length of the rectangle. However, the diagonals of the rhombus are equal to the length and width of the rectangle, making statement II true.

Statement III is true, because the area of the rhombus here is $\frac{diag1 \bullet diag2}{2}$, making the area of the rhombus equal to the half of the area of the rectangle. Statement IV is false, because the longest diagonal of the rhombus equals the length of the rectangle (which is less than the diagonal length of the rectangle).

Answer: C

33. In the following figure, the diagonal lines of the rectangle meet at the center of the circle. The rectangle inscribes the circle, and its dimensions are 4 feet and 7 feet. The chord of the circle measures *x* feet and is parallel to the length of the rectangle. What is the length of this chord (to the nearest 0.1) ?

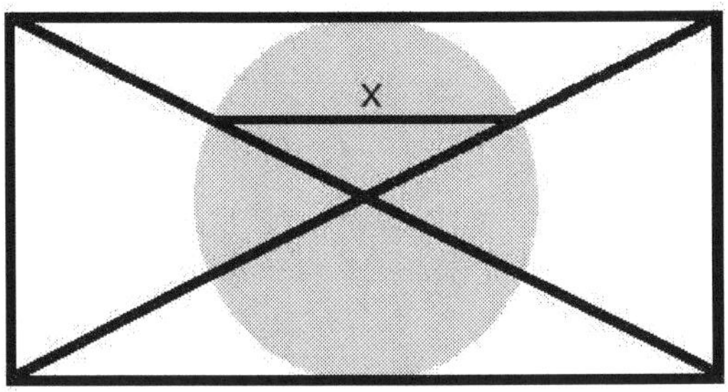

A. 3.5 ft

B. 3.2 ft

C. 2.9 ft

D. cannot be determined

This problem might seem more challenging than average geometry problems. We first start out with what we know: the width of the rectangle is 4, making the radius of the circle 2.

We must then throw out the circle and leave only the isosceles triangle with segment *x* in the picture (top portion of the rectangle). The following figure illustrates this.

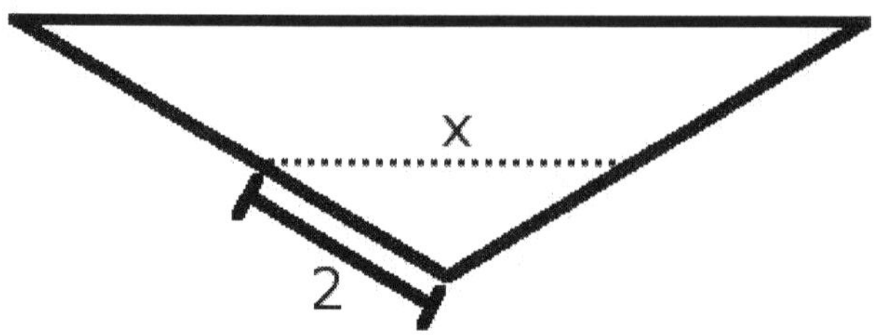

The base of this isosceles triangle is 7 (originally the length of the rectangle). We will divide the triangle into two equal right triangles (because it is isosceles), and any one of these right triangles (we can choose the left one to work with) will have sides $\frac{7}{2}$, 2 and hypotenuse 2 + *y*; this triangle has a small triangle inside it with sides $\frac{x}{2}$, 2 and an unknown third side. We first need to find the hypotenuse of the large triangle (2 + *y*). We use the *Pythagorean Theorem*: $(2 + y)^2 = 3.5^2 + 2^2$, and solving for *y* we get *y* = 2.031128874, so that the large triangle hypotenuse is 2 + 2.031128874 = 4.031128874. Now we can set up a direct proportion (because the 3 sides show proportional decrease when compared to the large triangle) to find *x*: we have $\frac{\frac{x}{2}}{2} = \frac{3.5}{4.031128874}$. Solving for *x*, we get *x* = 3.472972569.

Answer: A

34. In the following figure, the equation of the circle is $(x + 4)^2 + (y - 4)^2 = 8$. The smallest principal radius of the shaded circle region is 1.25 meters. What is the area of the non-shaded circle region (to the nearest 0.01)?

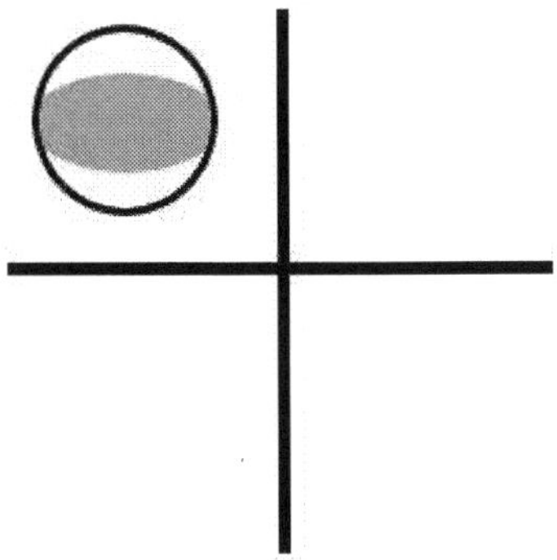

A. 11.77 m^2

B. 15.06 m^2

C. 12.39 m^2

D. 14.03 m^2

The non-shaded circle region is an enclosed ellipse, with its longest principal radius equal to the radius of the circle. The radius of the circle is calculated using a given circle equation, where 8 (on the right side of the equation) corresponds to the radius squared, giving $r = \sqrt{8}$.

We must subtract the area of the ellipse from the area of the circle to get the desired non-shaded circle region. Ellipse area is $\pi(1.25)(\sqrt{8}) = 11.10720735$. Area of the circle is 8π. Thus, the desired non-shaded area is $8\pi - 11.10720735 \approx 14.03$ square meters.

Answer: D

35. In the following figure, a rectangle is shown on the Cartesian plane. Two coordinates of its corners are also shown. The origin is the center point of the rectangle. What is the area of the shaded rectangle region (to the nearest hundredth)?

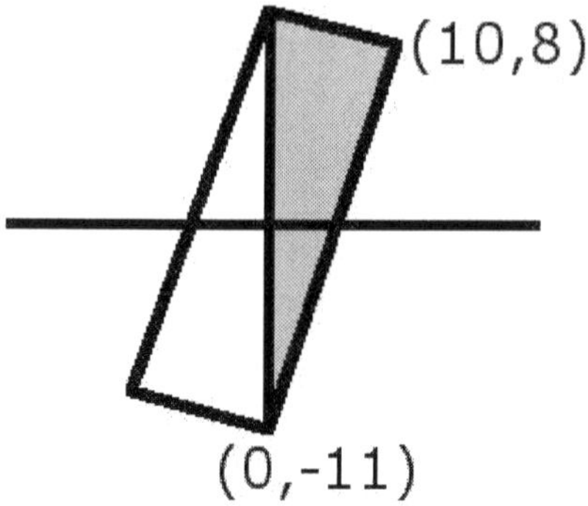

A. 95.12

B. 202.18

C. 101.09

D. 50.54

Since the hypotenuse of the shaded triangle in question is 22 (using coordinates of the two points joining the diagonal), we now must find the shortest triangle side. This is done by *Pythagorean Theorem* which will use the imaginary horizontal distance covered by this side along with the vertical distance covered by it (creating a small triangle with a hypotenuse that is the shortest side of the big shaded triangle in question, and is also the width of the rectangle shown).

Horizontal distance is 10 (using coordinates), and vertical distance is 3 (again using coordinate logic). Now, the small triangle hypotenuse using *P. T.* will be $\sqrt{10^2 + 3^2}$ = 10.44030651. This is the shortest side of the shaded triangle in question (also the width of the rectangle).

We now need to find the length of the rectangle to calculate the area of the shaded triangle in question. We use the *P. T.* again, getting length = $\sqrt{22^2 - 10.44030651^2}$ = 19.36491673. Thus, the area of the shaded region is $\frac{(19.36491673)(10.44030651)}{2}$ ≈ 101.09.

Answer: C

36. A rectangular solid has the following dimensions: height = 6 feet, width = 5 feet, and length = 11 feet. Water is poured into the solid, filling it $\frac{7}{11}$ of its volume. What is the water level (in inches) after the water has been filled (to the nearest 0.1)?

A. 3.8 inches

B. 4.2 inches

C. 45.8 inches

D. 50.4 inches

We are asked to find the height (the water level) after the water has filled $\frac{7}{11}$ of the volume of the rectangular solid. First, we must find the volume of the solid. This is done by the volume formula (*length* • *width* • *height*) = 6 • 5 • 11 = 330 cubic feet.

Now we find the volume of the water filled into the solid: $\frac{330(7)}{11}$ = 210 cubic feet. Thus, solving for the height we get: 210 = 11 • 5 • *height*, giving *height* = 3.818181818. We must remember to convert the answer to inches (as required by the problem). Therefore, the height becomes 3.818181818 • 12 ≈ 45.8 inches.

Answer: C

37. A triangular prism has the following dimensions: length = 12 feet, base = 7 feet, and height = 8 feet. Water is poured into the prism, taking $\frac{5}{9}$ of the prism volume. A small triangular prism is formed after the water has been filled (where there is no water). What is the ratio of the volume of the small unfilled prism formed to the volume of the original prism (to the nearest 0.01)?

A. 0.37

B. 0.53

C. 0.48

D. 0.44

We need to find the volume of the prism first: $\frac{bhw}{2} = \frac{12(7)(8)}{2}$ = 336. Now, we can find the small unfilled prism by knowing the fact that since the water takes $\frac{5}{9}$ of the large prism volume, the unfilled portion of the volume of the large prism will be $\frac{4}{9}$ of its volume. Thus, the unfilled portion of the volume is $\frac{336(4)}{9}$ = 149.3333333. Finally, the ratio of the unfilled prism volume to the original volume of the prism is $\frac{149.3333333}{336} \approx$ 0.44.

Answer: D

38. A triangle whose base is 13 meters and height 9 meters is painted so that the height of the painted region is 5 meters (see figure). What is the area of the non-painted region?

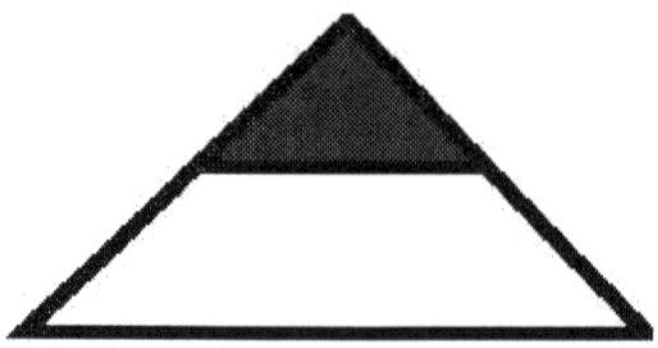

A. 40 square meters

B. 49 square meters

C. 51 square meters

D. 44 square meters

There are two ways of solving this problem: the trapezoid way and the triangle way. The trapezoid way involves noticing that the non-painted region is a trapezoid, whose height is 9 – 5 = 4 meters.

To find the unknown base of the trapezoid (also the base of the painted small triangle), we can use a direct proportion, because the small triangle shows proportional decrease on all sides with respect to the large triangle. We will then have: $(\frac{5}{9}) = (\frac{x}{13})$, and our x (the unknown base of the small triangle) is 7.222222222. We could have set up a different direct proportion: $(\frac{5}{x}) = (\frac{9}{13})$, and we would still get the right answer. Note: it does not matter how you choose to set up the direct proportion, as long as you are consistent with what is on the numerator and denominator in the fractions used. This is basically why the two proportions worked to find the missing x.

Now, since we found the base of the triangle, we can use it to find the area of the trapezoid (the non-painted part of the large triangle). We have $\frac{(b_1 + b_2)h}{2} = \frac{(7.222222222 + 13)4}{2} \approx 40$ square meters. The triangle way of solving this problem involves subtracting the area of the painted region (small triangle) from the area of the large triangle. The area of the small triangle is $\frac{(7.222222222)(5)}{2} = 18.05555556$. The area of the large triangle is $\frac{(13)(9)}{2} = 58.5$. Thus, the desired area of the non-painted triangle region is 58.5 – 18.05555556 ≈ 40 square meters.

Answer: A

39. In the following figure, an enclosed square shares the center point with the circle. If the area of the square is 13, what is the area of the circle region excluding the square?

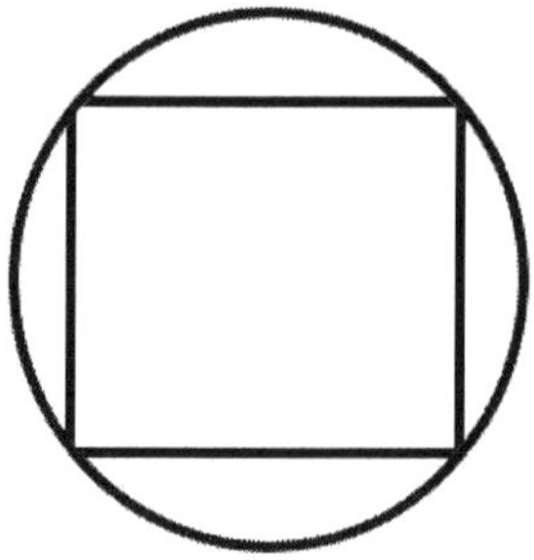

A. 13π – 13

B. $\frac{169\pi}{2}$ – 13

C. 169π – 13

D. $\frac{13\pi}{2}$ – 13

We must find the diameter of the circle, which is the diagonal of the square. To do this, we use the *Pythagorean Theorem*: diameter = $\sqrt{(\sqrt{13^2}) + (\sqrt{13^2})} = \sqrt{13 + 13} = \sqrt{26}$, which means the radius of the circle is $\frac{\sqrt{26}}{2}$. Thus, the desired area of the circle region excluding the square region is $\pi r^2 - 13 = \frac{13\pi}{2} - 13$.

Answer: D

40. The following figure shows a square whose small squares have been cut out at its corners. Each small square side is $\frac{1}{19}$ of the perimeter of the large square. If the perimeter of the large square is 32, what is the area of the figure without the small squares?

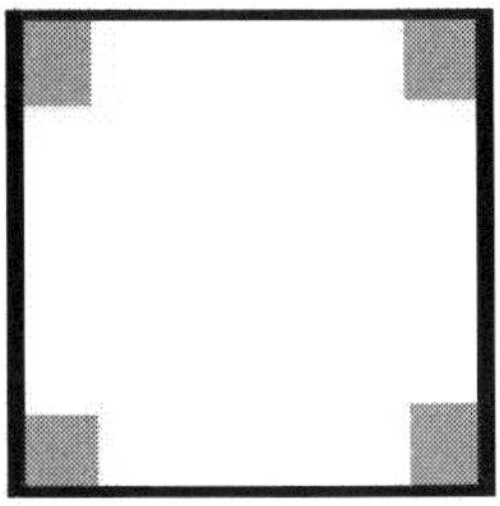

A. 53

B. 61

C. 49

D. 57

There are two ways of solving this problem: subtracting the areas of the small squares from the area of the large square, or adding the areas of the small rectangles formed between the adjacent small squares with the area of the square formed by margins created by the small squares.

The first way is the easiest way: each small square side is $\frac{1}{19}$ of the perimeter of the large square, so that if the perimeter is 32, then the small square side is $\frac{32}{19}$. The area of the small square is then $(\frac{32}{19})^2$ = 2.836565097. Since the perimeter of the large square is 32, each side of it must be 8. Thus, the desired area of the altered large square becomes (8 • 8) – 4(2.836565097) = 52.65373961. Now, the second way of doing this: we wish to find the large chunk of each side of the large square that does not include the small square sides. We will name this chunk *S*. Then, we will add the four *S* chunks with the small two sides and set the sum equal to 32 (the perimeter). We get 32 = $4S + (2)(\frac{32}{19})(4)$, and we get *S* = 4.631578948.

There are four small rectangles formed between the adjacent small squares. The area of each small rectangle is then $4.631578948 \cdot (\frac{32}{19}) = 7.800554018$. Now, a square is formed (which shares the same center with the original large square) by the margins created by the small squares, and its side is the same as the S chunk we found: 4.631578948. The area of this square is $4.6315789482^2 = 21.45152355$. All we need to do now is to add the areas of the rectangles with the area of the square formed by the margins. This becomes $(4 \cdot 7.800554018) + 21.45152355 = 52.65373962$.

Answer: A

41. The following information is thrown regarding the ratios of sides for the right triangle:

I. The ratio of 5:12:13 is possible

II. The ratio of 8:15:17 is possible

III. the ratio of 7:24:25 is possible

IV. the ratio of 3:4:5 is possible

Which statements are true?

A. all of the above

B. none of the above

C. I and II only

D. I, II and IV only

All ratios in the right triangle must satisfy the properties of the *Pythagorean Theorem.* Before you check that, however, be sure to check whether the ratios satisfy the properties of any triangle: the difference between any two sides must be less than the third side, and the sum of any two sides must be greater than the third side. All ratios satisfy this requirement. Checking each statement now with *P. T.*, it happens that all of them are correct. Remember to check by putting the longest side (hypotenuse) on one side of the equation, and the two remaining sides on the other, using expression $c^2 = a^2 + b^2$.

Answer: A

42. The following statements are thrown about sides of a triangle:

I. the ratio of 1:1:$\sqrt{2}$ gives an isosceles triangle

II. The ratio of 1:2:$\sqrt{3}$ gives an equilateral triangle

III. The ratio of 1:2:$\sqrt{3}$ is a 45-45-90 degree triangle

IV. The ratio of 1:1:$\sqrt{2}$ is a 30-60-90 degree triangle

Which of the statements are FALSE?

A. I and IV only

B. I, II and III only

C. II and III only

D. II, III and IV only

Statement I is correct, because the ratio suggests that two of the three sides are equal. Statement II is incorrect for this same reason. Statement III is incorrect, because the angles must be proportional to the sides, and in this case two equal angles would mean that we would have two equal sides (which is not the case). Statement IV is incorrect for this same reason.

Answer: D

43. In the following large isosceles triangle, a small isosceles triangle is enclosed with two congruent sides, each 2 feet long. What is the measure of angle *ABC* and value of segment *AB*?

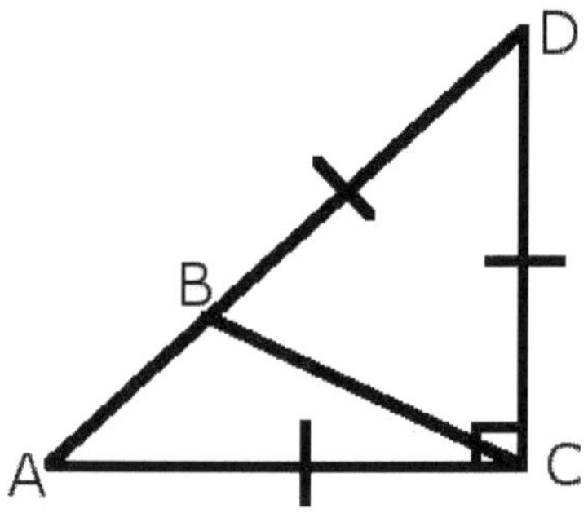

A. m<*ABC* = 112.5 degrees, *AB* = $(2\sqrt{2} - 2)$ feet

B. m<*ABC* = 145 degrees, *AB* = $2\sqrt{2}$ feet

C. m<*ABC* = 112.5 degrees, *AB* cannot be determined

D. m<*ABC* cannot be determined, *AB* = $(2\sqrt{2} - 2)$ feet

Since the large isosceles triangle is a right triangle (a 45-45-90 triangle), the two other angles (A and D) must measure 45 degrees each. The small isosceles triangle formed (triangle BCD) shares the 45-degree angle D. The congruent sides mean that the opposite angles to these sides must be also equal. Therefore, angles CBD and BCD are equal, and their sum will measure (180 – 45) = 135 degrees. Since they are equal, each will measure $\frac{135}{2}$ = 67.5 degrees.

Now, angle ABC must be a supplement of one of these angles, so it measures (180 – 67.5) = 112.5 degrees. To find segment AB, we must first understand the ratio of the sides of a right isosceles triangle: $1:1:\sqrt{2}$. Since AC = CD = 2, the third side AD must be multiplied by $\sqrt{2}$ to satisfy the ratio. So AD = $2\sqrt{2}$. Since BD = 2, AB = $2\sqrt{2} - 2$ feet long.

Answer: A

44. In the following figure, segment *BA* bisects angle *A*. Segment *EB* is $\frac{3}{2}$ in value. What is the sum of segments *BC* and *DE*?

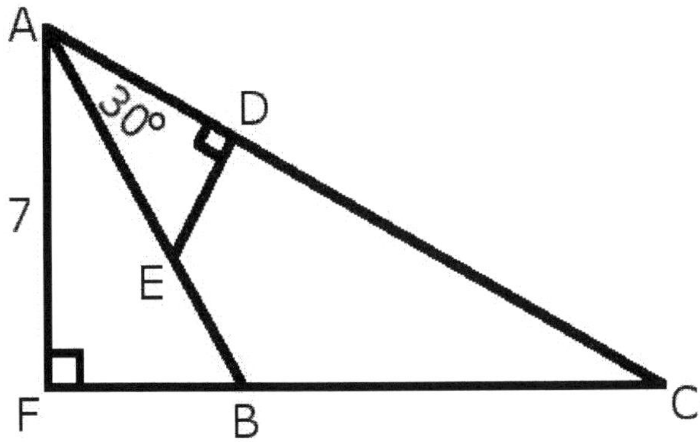

A. $\frac{21\sqrt{3}}{3}+\frac{3}{4}$

B. $\frac{21\sqrt{3}}{3}-\frac{3}{4}$

C. $\frac{14\sqrt{3}}{3}-\frac{3}{4}$

D. cannot be determined

This problem simply asks to use the ratio of the two 30-60-90 triangles to find their respective sides as well as their sum. Let's find segment BC first. Since segment BA bisects angle A, triangle FAC must be a 30-60-90 triangle. The ratio of the sides of the 30-60-90 triangle is 1:$\sqrt{3}$:2. Since AF is 7, FC must be $7\sqrt{3}$.

Now, since 7 is not the shortest side of triangle FAB, FB must be $\frac{7}{\sqrt{3}}$. Thus, BC = FC – FB = $7\sqrt{3}-\frac{7}{\sqrt{3}}=\frac{14\sqrt{3}}{3}$. Now we will find DE. Since AF is 7, AB must be $\frac{14}{\sqrt{3}}$, making AE = AB – EB = $\frac{14}{\sqrt{3}}-\frac{3}{2}$. Since triangle ADE is a 30-60-90 triangle, the shortest side is

DE (what we need to find), and it must be $\frac{\frac{14}{\sqrt{3}}-\frac{3}{2}}{2}$ since AE = 2 • DE. Thus, our desired sum is DE + BC = $\frac{\frac{14}{\sqrt{3}}-\frac{3}{2}}{2}+\frac{14\sqrt{3}}{3}$, or $\frac{21\sqrt{3}}{3}-\frac{3}{4}$.

Answer: B

45. The following figure shows a rhombus with its intersecting diagonals. If the half of the longest diagonal is $\sqrt{3}$ feet, what is the area (in inches) of the rhombus?

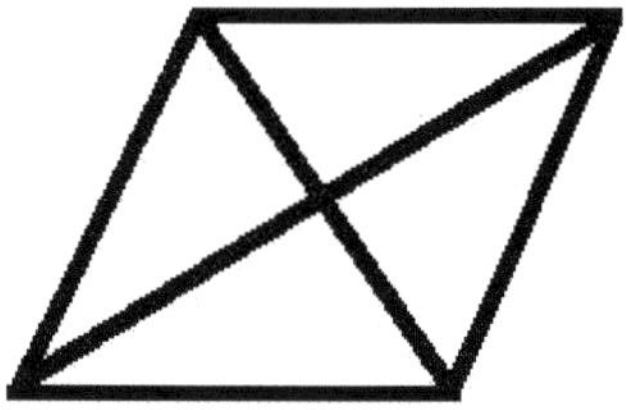

A. $2\sqrt{3}$ in ^{2}

B. $24\sqrt{3}$ in ^{2}

C. $288\sqrt{3}$ in ^{2}

D. $144\sqrt{3}$ in ^{2}

Since the diagonals of the rhombus are perpendicular to each other, the four central angles are each 90 degrees in measure. Now, we are given $\sqrt{3}$ as the half of the longest diagonal. If we divide the rhombus into four equal triangles, we can see that the half of the longest rhombus diagonal is a side shared by two of the four small triangles.

Using the fact that one of the sides of these triangles is $\sqrt{3}$, we can clearly see that the triangle with a 90 degree angle and adjacent side $\sqrt{3}$ is a 30-60-90 triangle. In fact, all of the four equal triangles are 30-60-90 triangles. With this information, we can find the other diagonal. First we find the half of the shorter diagonal. The short diagonal of the triangle we worked with is the smallest side of that triangle, and since we know that a 30-60-90 triangle has side ratio 1:$\sqrt{3}$:2, the shortest side is 1. This is the half of the shorter diagonal. So this makes the shorter diagonal 1 • 2 = 2.

Now, the area of the rhombus is determined by multiplying the longest diagonal with the shorter diagonal, and then dividing the product by 2. Our longer diagonal is $\sqrt{3} \cdot 2 = 2\sqrt{3}$. Thus, the area of the rhombus is $\frac{2(2)\sqrt{3}}{2} = 2\sqrt{3}$. Converting the answer to inches, we get $2\sqrt{3} \cdot 144 = 288\sqrt{3}$ square inches.

Answer: C

46. In a trapezoid, if you increase one base by 2, increase the other base by 2, and decrease the height by 4, what is the percent increase/decrease of the area of the newly created trapezoid with respect to the area of the original trapezoid?

A. new area increases by 50%

B. answers vary

C. the two areas are the same

D. new area decreases by 150%

We need to test different numbers to see if this can be disproven as a general formula. If we take h = 6, $b1$ = 10, $b2$ = 10, our area will be 60. Then, our new area is calculated using h = 2, $b1$ = 12, and $b2$ = 12, and it is 24. The area ratio is $(\frac{24}{60}) = 0.4$. Now, if we plug in the original numbers as h = 6, $b1$ = 1, and $b2$ = 1 and compare the areas, we get both areas to be 6. The area ratio now will be 1. Therefore, our answer depends on the chosen dimensions.

Answer: B

47. If a rhombus side measures 12 feet, and its smallest angle measures 60 degrees, what is the area (in square yards) of the rhombus (to the nearest tenth)?

A. 13.9 square yards

B. 125.1 square yards

C. 62.4 square yards

D. 6.9 square yards

Any rhombus will have an acute angle between its base and slanted height. In this example, its smallest acute angle measures 60 degrees. We can clearly see that a 30-60-90 triangle forms by the boundaries of the slanted height (hypotenuse), vertical height and the base of the rhombus. Since the ratio of sides for 30-60-90 triangle are $1:\sqrt{3}:2$, and the hypotenuse is 12 (the largest side corresponding to 2 in the side ratio), our remaining two sides must be 6 (base) and $6\sqrt{3}$ (height). Our area is then *base* • *height* = $6 \cdot 6\sqrt{3} = 36\sqrt{3}$ square feet. Converting to yards (where 1 square yard = 9 square feet), we get the area to be $\frac{36\sqrt{3}}{9} = 4\sqrt{3} \approx 6.9$ square yards.

Answer: D

48. In the following figure, if the area of the polygon is 64 square feet, what is the value of its diagonal?

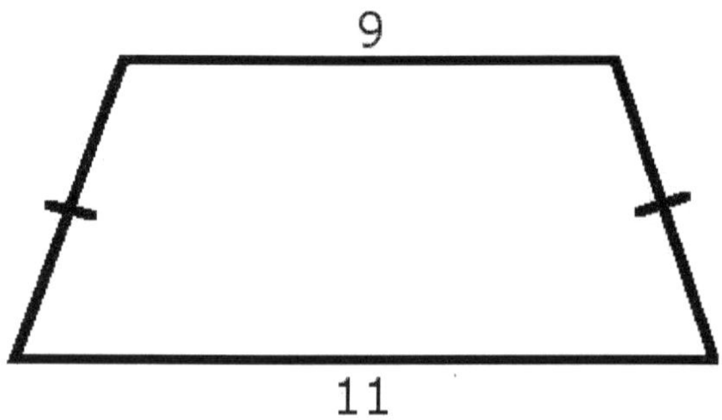

A. 11 feet

B. 10 feet

C. 12 feet

D. 13 feet

First, we solve for the height: $64 = \frac{(9+11)h}{2}$, getting $h = 6.4$. Before using the *P. T.* to find the diagonal (hypotenuse), we must notice that the base to be used for the *P. T.* formula will be 10 (not 11), because the diagonal spans a horizontal distance of 10. Our diagonal is $\sqrt{10^2 + 6.4^2} \approx 12$ feet.

Answer: C

49. In the following figure, all sides are equal. A line *x* connects the two corners of the polygon. What is the length of the line?

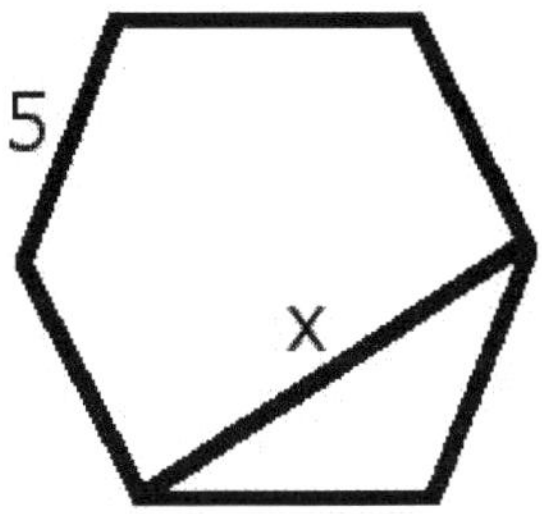

A. $5\sqrt{3}$

B. 10

C. $\frac{5\sqrt{3}}{2}$

D. 8

This is an equilateral hexagon (6 identical sides), with all interior angle measure summing to $180(s - 2) = 180(6 - 2) = 720$ degrees, and each interior angle measure is $\frac{720}{6} = 120$ degrees.

If we look at the isosceles triangle bounded by the line *x*, we can see that this is a 120-30-30 triangle made up two of identical 30-60-90 triangles (if we divide the isosceles triangle in half). The side ratio of the 30-60-90 triangle is $1:\sqrt{3}:2$, and since the hypotenuse is 5, the side we are looking for is half of line *x*, and corresponds to $\sqrt{3}$ in the side ratio. So, it is $\frac{5\sqrt{3}}{2}$. Thus, line *x* is twice this length, namely $5\sqrt{3}$.

Answer: A

50. In the following figure, the height of the closed right cylinder and one leg of an enclosed right triangle match in length. If the surface area of the cylinder is 49 square feet, what is the diameter of the cylinder?

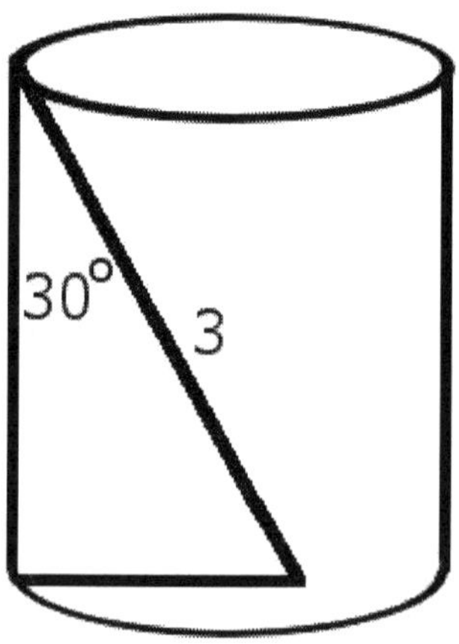

A. 5 feet

B. 2 feet

C. 3 feet

D. 4 feet

To use the formula for the surface area, we must first find the height of the cylinder, which is the adjacent side relative to the hypotenuse of the triangle. Since this is a 30-60-90 triangle (with side ratio 1:$\sqrt{3}$:2), the adjacent side will be $\frac{3\sqrt{3}}{2}$.

This is the height of the cylinder. Now we can use the surface area formula and solve for the radius of the cylinder. We have $49 = 2\pi r^2 + 3\sqrt{3}\pi r$, and solving for r using the quadratic formula, we get r = 1.780911929 (we reject the negative root), and the diameter is twice this amount, that is, 3.561823858.

Answer: D

51. If the dimensions of a rectangular prism are 3.5 feet, $\frac{11}{4}$ feet and 11 feet, what is the value of the longest diagonal of the prism?

A. 12 feet

B. 9 feet

C. 11 feet

D. 10 feet

It does not matter what dimensions we choose to assign to the given numbers. Essentially, we simply have to find the longest diagonal that will connect the farthest corners of the prism. It will be 3-dimensional. There are four such diagonals (all identical). To find any one of them, we must use the *P. T.* by simply extending it (adding the third squared dimension to the general formula). We will have $d^2 = a^2 + b^2 + c^2$, where *a, b,* and *c* will be our given dimensions. So, $d = \sqrt{3.5^2 + \left(\frac{11}{4}\right)^2 + 11^2} \approx 12$ feet.

Answer: A

52. If the diameter of a base of a right circular cone measures 12 meters, and the volume of a cone is 154 cubic meters, what is the slant height of the cone?

A. 5 meters

B. 6 meters

C. 7 meters

D. 8 meters

Since the radius of the base of a cone is 6, solving for the height using cone volume formula V = $\frac{\pi r^2 h}{3}$, that is, 154 = $\frac{36\pi h}{3}$, we get h = 4.084976873. We now use the *P. T.* to find the slanted height (radius is the opposite side of the slant height, vertical height is the adjacent side). So, our slant height is $\sqrt{4.084976873^2 + 36}$ ≈ 7 m.

Answer: C

53. If the slant height of a right circular cone measures 10 feet, and the angle between the slant height and vertical height is 30 degrees, find the surface area of the cone.

A. $(75 + 50\sqrt{3})\pi$ ft ^{2}

B. $(\frac{125\pi}{2})$ ft ^{2}

C. $(\frac{125\pi}{4})$ ft ^{2}

D. 75π ft ^{2}

Since the angle between the slanted height and the vertical height of a right cone measures 30 degrees, this is a 30-60-90 triangle we can use for reference to find the base radius and vertical height. Knowing the side ratio for the 30-60-90 triangle, we see that the base radius is 5, and vertical height is $5\sqrt{3}$. We can now use this information to find the surface area of a cone. Using the formula, we get $S_A = \pi r(r + \sqrt{r^2 + h^2}) = 5\pi(5 + \sqrt{25 + 75}) = 75\pi$ square feet.

Answer: D

54. If a right rectangular pyramid has identical isosceles triangle faces, with each face having a 7-foot base, and pyramid's longest slanted height 6 feet, find the surface area of the pyramid.

A. 133 square feet

B. 83 square feet

C. 117 square feet

D. 91 square feet

Since the length and width of the pyramid are the same, measuring 7 feet (square base of a pyramid is given), all we now need is the vertical height to use the surface area formula. This formula works because the center point of the base face is connected by a perpendicular line (to the base face) to the tip (apex) of the pyramid.

We are also given the longest slanted height as 6 (this is a slanted height serving as a boundary for any two adjacent triangle faces of a pyramid), the shortest slanted height to the tip (apex) of a pyramid will be the height in the center of each triangle face extending from the base to the tip of the pyramid (with base length now covering half the distance of 7, or 3.5).

To find this shortest slanted height, we use the *P. T.*: $\sqrt{6^2 - 3.5^2} = 4.873397172$. We can now use this height to find our vertical height: $h = \sqrt{4.873397172^2 - 3.5^2} = 3.391164992$. Thus, using the surface area formula $lw + w\sqrt{(\frac{l}{2})^2 + h^2} + l\sqrt{(\frac{w}{2})^2 + h^2}$ and plugging in the dimensions, we get $S_A \approx 117$ square feet.

Answer: C

55. In a trapezoidal prism, there is only one acute angle between the two faces, which is 60 degrees. If the prism's long base is 11.5 centimeters, slanted height 4.5 centimeters, and length $\frac{33}{2}$ centimeters, find the volume (in cubic inches) and surface area (in square inches, to the nearest tenth) of the trapezoidal prism.

A. V = 87.1 in ^{3}, S_A = 40.7 in ^{2}

B. V = 40.7 in ^{3}, S_A = 87.1 in ^{2}

C. V = 91.2 in ^{3}, S_A = 36.8 in ^{2}

D. V = 36.8 in ^{3}, S_A = 91.2 in ^{2}

Since there is only one acute angle between the faces of this trapezoidal prism, we can think of a hybrid figure of a rectangle joined with a right triangle, creating a trapezoid. See the following figure of what this 2-dimensional shot may look like:

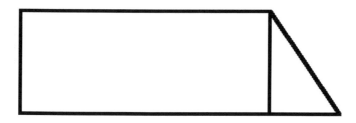

Of course, we extend this figure with a length to create a prism. Now, the vertical height is imaginary, so the only acute angle is indeed between the bottom base (longest base) and the slanted height, which is given in measure as 60 degrees. To find the top base (shorter base), we subtract the adjacent side of the 60-degree angle from the total bottom base distance. We use the 30-60-90 triangle side ratio for this: since the slanted height h_s is 4.5, the adjacent side is 2.25. So, our top base is 11.5 – 2.25 = 9.25. Vertical height is $2.25\sqrt{3}$ (again using side ratio).

The surface area then is $2(b1 + b2)(\frac{h}{2}) + l(b1 + b2 + h + h_s) = 561.7925083$. Converting to square inches (where 1 square inch = 6.4516 square cm), we get $(\frac{561.7925083}{6.4516}) \approx$

87.1 square inches. The volume of the prism is $(b1 + b2)(\frac{h}{2})(l)$ = 667.137257 cubic cm. Converting to cubic inches (where 1 cubic inch = 16.387064 cubic cm), we get $(\frac{667.137257}{16.387064}) \approx 40.7$ cubic inches.

Answer: B

56. In the following figure, if the height of the triangle *BCD* is 4 feet, segment *BC* is 11 feet, and measure of angle *DAB* is 30 degrees, what is the area and perimeter of triangle *ABCD*?

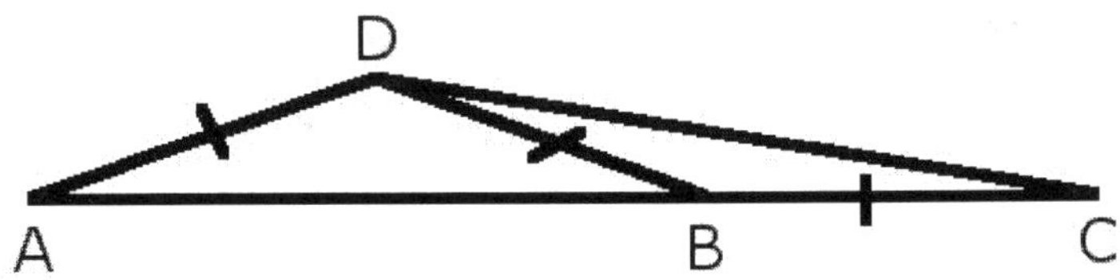

A. A = 62 square feet, P = 93 feet

B. A = 8,928 square feet, P = 1,116 feet

C. A = 1,116 square feet, P = 8,928 feet

D. A = 93 square feet, P = 62 feet

Since the triangle BCD is isosceles, height from segment CD to angle B will divide CD in half (and the isosceles triangle into two identical right triangles). We find this CD half (adjacent side for one of the small right triangles created) by using the *P. T.*: $\sqrt{11^2 - 4^2}$ = 10.24695077. This makes segment CD twice this amount, or 20.49390154. Now, since AD = BC = DB, triangle ADB is also isosceles, and we can divide this triangle in half to find its height (joining the midpoint of segment AB with angle D).

We now use the fact that there is a 30 degree angle in the left small right triangle created (after we divided triangle ADB in half), so we use the 30-60-90 degree side ratio to find the height. Since AD = 11, height must be $\frac{11}{2}$, or 5.5. Half of AB is then 9.526279442 (this we get from either *P. T.* or side ratio for adjacent). AB is twice this amount, that is, 19.05255888. Thus, the area of the entire triangle ABCD is made up of combined areas of triangles ABD and BCD, that is $\frac{(19.05255888)(5.5)}{2}$ + $\frac{(20.49390154)(4)}{2}$ ≈ 93 square feet. Perimeter is 2(11) + 19.05255888 + 20.49390154 ≈ 62 feet.

Answer: D

57. In the following figure, if the central acute angle *AFD* measures 84 degrees, and inscribed angle *ABC* measures 80 degrees, what is the measure of major arc *DC*?

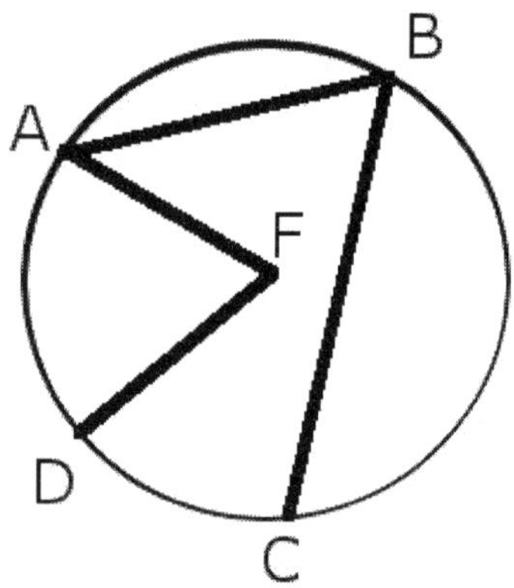

A. 196 degrees

B. 160 degrees

C. 284 degrees

D. 200 degrees

Since angle ABC measures 80 degrees, minor arc AC measures twice this amount, or 160 degrees. Minor arc AD is exactly what the central acute angle F is in measure, that is, 84 degrees.

This means that minor arc DC is (minor arc AD – minor arc AC) = 160 – 84 = 76 degrees in measure, which means that the major arc DC will be 360 – 76 = 284 degrees in measure.

Answer: C

58. In the following figure, chord *AD* intersects with segment *EF* at circle center E. Chord *BC* || *EF*. If the area of the circle is 196π, and major arcs *CD* and *AB* measure 270 and 330 degrees, what is the value of segment *EF*?

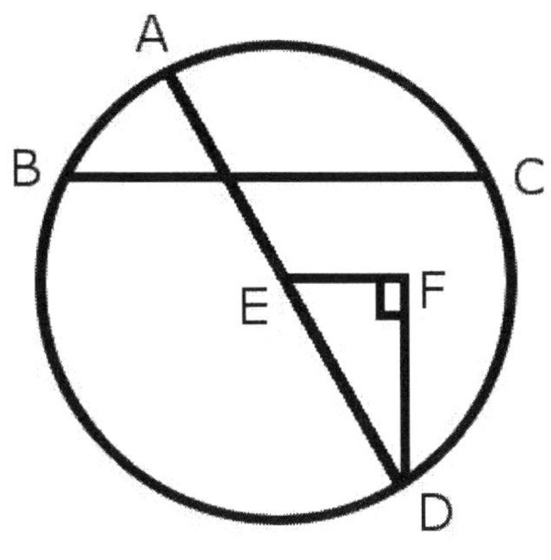

A. $\sqrt{42}$

B. $14\sqrt{3}$

C. 7

D. $7\sqrt{3}$

We first find the radius of the circle: $196\pi = \pi r^2$, which means that the radius is 14. This is the hypotenuse of triangle EFD. We know that since segments EF and BC are parallel, the two vertical angles created by two intersecting chords AD and BC will each be equal in measure to angle FED. We find this angle measure as follows.

We know that the angle created by two intersecting chords in a circle will be one-half the sum of intercepted arcs in degrees, so we find the minor arcs CD and AB. Major arc CD measures 270 degrees, so minor arc CD will measure 360 – 270 = 90 degrees. Major arc AB measures 330 degrees, so minor arc AB measures 30 degrees. Now we can find the vertical angle measure (which is also angle FED): $\frac{90 + 30}{2}$ = 60 degrees. This is the measure of angle FED. Now, we see this is a 30-60-90 triangle, so we use the side ratio to find adjacent side EF: since the radius is 14 (hypotenuse), EF is 7.

Answer: C

59. In the following figure, major arc *BD* measures 340°, and minor arc *AC* measures 70°. Segments *BE* and *DE* are equal. If segment *BE* is 2 feet, and chord *CD* is 3.5 feet, what is the height and area of triangle *CDEBA* ? (note: figure not drawn to scale.)

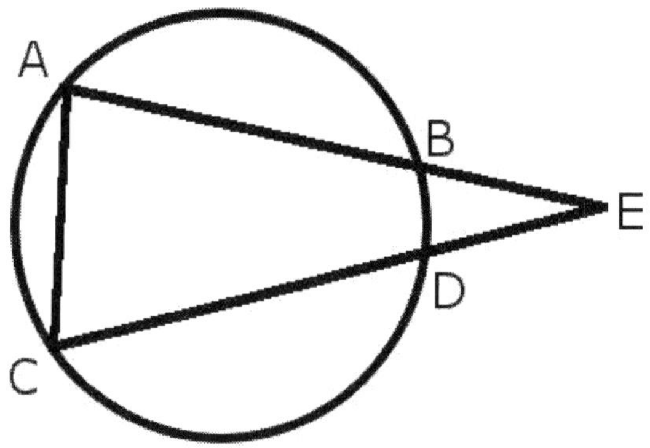

A. h = 6 ft, A = 5 ft ^{2}

B. h = 5 ft, A = 6 ft ^{2}

C. h = 3 ft, A = 5 ft ^{2}

D. h = 5 ft, A = 3 ft ^{2}

Here we use the fact that the angle outside the circle created by two tangents, two secants or combination of a secant and a tangent is equal in degrees to one-half the difference of the intercepted arcs. In this example, two secants create angle E. So we find the intercepted arc measures first: minor arc BD measures 360 – 340 = 20 degrees, and minor arc AC measures 70 degrees. Angle E then measures $\frac{70 - 20}{2} = 25$ degrees. Since BE = DE, and points A and C lie on the circle, AE = CE = 2 + 3.5 = 5.5. This makes the triangle isosceles, so we can divide it in half to find its base and height, and the area.

Angle E becomes 12.5 degrees for each of the small right triangles. We cannot use the side ratio, because it is not a 30-60-90 triangle, but we can use trigonometry to find the base and height *h*. Since $\cos(12.5^{\circ}) = \frac{h}{5.5}$, *h* = 5.369628039 ft. Base *b* is found in a similar way: since $\sin(12.5^{\circ}) = \frac{b}{5.5}$, *b* = 1.190417877. The large base *B* (for the original large triangle in the figure) is then twice this amount, or 2.380835753. The area is then $\frac{Bh}{2}$ = 6.392101209 square feet.

Answer: B

60. In the following figure, three circles are inscribed in an isosceles triangle *ABC* whose sides are tangent to the circles at the indicated points. Circles are also tangent to each other, the large circle shares the center point with the triangle, and two small circles are identical. *AD* = 2 meters, *GC* = 3.5 meters, and *AB* = 9.5 meters. What is the height and area of the triangle (in feet)?

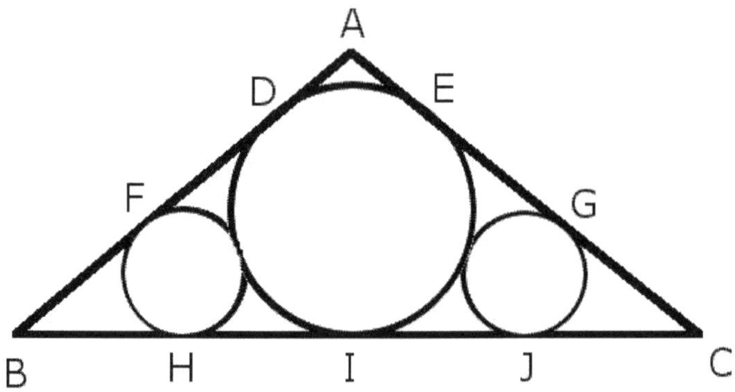

A. 6 feet, 471 square feet

B. 19 feet, 44 square feet

C. 6 feet, 44 square feet

D. 19 feet, 471 square feet

Since AC = AB, and all points are tangent on the similar circles (and the large circle), FD = EG = AB – BF – AD = 4. Now, since BH and BF are both tangent to the same small circle, they are equal to each other. Moreover, since BD and BI are both tangent to the large circle, they are also equal, which means that FD = HI = 4. This makes BI = 3.5 + 4 = 7.5. Because the large circle shares the center point with the original large triangle, BI is exactly half of the triangle base BC.

Knowing that AB = 9.5, we can calculate the height *h* by dividing the isosceles triangle in half and using the *P. T.*: $\sqrt{AB^2 - BI^2}$ = 5.830951895 m. Converting to feet (where 1 m = 3.280839895 ft), height is about 19 feet. Area in meters is $\frac{(15)(5.830951895)}{2}$ = 43.73213921 m^2. Knowing that 1 square meter = 10.76391042 square feet, our answer becomes about 471 square feet.

Answer: D

61. If a torus has an inner radius that is the same length as one of the identical sides of the referenced 45-45-90 triangle whose height drawn from the longest

side (hypotenuse) is 5.5 meters, and the outer radius of the torus is 2.5 meters, find the volume (in cubic feet) of the torus.

A. 960 ft^3

B. 58,558,008 ft^3

C. 1,000,000 ft^3

D. 33,888 ft^3

Here we must be careful to understand which height of the isosceles triangle the problem talks about. This depends on how we draw the isosceles triangle. The correct way to do this is to have the longest side (the rest two are shorter and identical) of the triangle lie as a base. If we were to assign one of the identical sides to be the base of the triangle the height would be of different length. See figure below for comparison. These are identical 45-45-90 triangles that simply vary in their position. However, their heights are different. The first figure is what the problem asks.

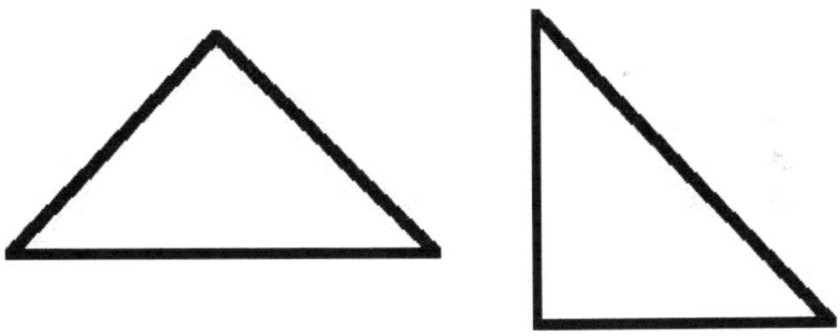

Now, since the height in the left figure is 5.5, half of the base will also be 5.5 (why?). This makes a small 45-45-90 triangle. The large radius in question is the hypotenuse of this small triangle that we need (which is one of the identical sides of the original large triangle). We use the *P.T.* to find it, getting 7.778174593. This is the large radius R of the torus. Our small radius r is 2.5. Now we can find the volume of the torus: $V = (\pi r^2)(2\pi R) = 959.5938274$ cubic meters. Converting to cubic feet (1 cubic meter = 35.3147 cubic feet), we get about 33,888 cubic feet.

Answer: D

62. The following figure shows a kite enclosed in a rectangle whose dimensions are 6 feet and 13 feet. What is the sum of perimeters (in meters) of the rectangle and kite?

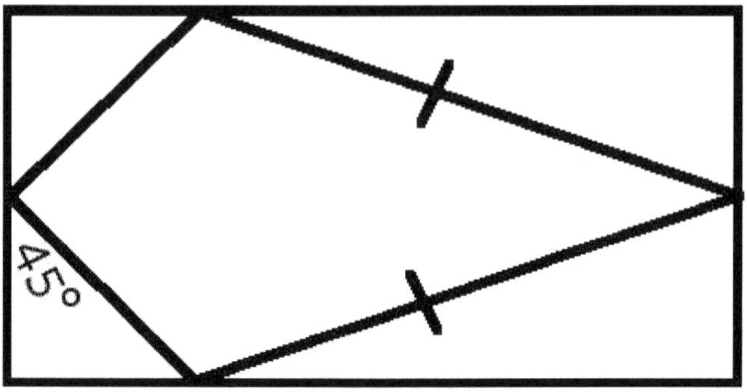

A. (38)(0.3048) m

B. $(2\sqrt{109} + 6\sqrt{2})(0.3048)$ m

C. $(38 + 2\sqrt{109} + 6\sqrt{2})(0.3048)$ m

D. $(26 + \sqrt{109} + \sqrt{2})(0.3048)$ m

Perimeter of the rectangle is 2(6) + 2(13) = 38 feet. Looking at the two short identical sides of a kite, it is easy to see that they create two small isosceles triangles whose base and height takes a small part of the rectangle length and width. This is known because one of the angles measures 45 degrees, making the opposite angle 45 degrees as well (since the third angle automatically measures 90 degrees because it is an edge of a rectangle).

Since we have two 45-degree angles, their respective opposite sides are equal in length (base and height of the small isosceles triangles). Their length must be 3 feet, because the two identical sides meet in the middle of the rectangle width (due to 45 degrees between each of the both small hypotenuses and the rectangle width, and because hypotenuses are equal).

Each of the small hypotenuses (longest side of each small isosceles triangle) is $\sqrt{3^2+3^2}$ = $3\sqrt{2}$ feet. Since part of the rectangle length is also 3 feet (base for each of the two small isosceles triangles), the rest of the rectangle length must be 13 – 3 = 10 feet. There are now two bigger identical right triangles, each with base 10, height 3 and respective hypotenuse. Each hypotenuse is then $\sqrt{10^2+3^2}$ = $\sqrt{109}$. So, our desired combined perimeter for both the kite and rectangle is $2\sqrt{109}$ + $2(3\sqrt{2})$ + 38 = 38 + $6\sqrt{2}$ + $2\sqrt{109}$ feet. All we have to do now is to note that 1 foot = 0.3048 meters, so we multiply our combined perimeter by 0.3048 to convert our answer to meters.

Answer: C

63. The following figure shows a 3-dimensional rectangular coordinate figure. The coordinates of the bottom four corner edges are (-1,4,3), (-5,4,3), (-1,11,3) and (-5,11,3). The given coordinate system is measured in feet. If the volume of the figure is 252 square feet, find the coordinates of the top four edges with respect to the inch coordinate system.

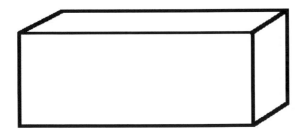

A. (-1,4,12), (-5,4,12), (-1,11,12) and (-5,11,12)

B. (-12,48,144), (-60,48,144), (-12,132,144) and (-60,132,144)

C. (-1,4,12), (-5,4,12), (-12,132,144) and (-60,132,144)

D. (-12,48,144), (-60,48,144), (-1,11,12) and (-5,11,12)

We are asked to find the coordinates of the top four corner edges as if we were setting up an inch coordinate window. This means that we must also convert all the known coordinates into inches before we decide on the unknown coordinates. The known set of coordinates is (-1,4,3), (-5,4,3), (-1,11,3) and (-5,11,3). These are given in feet. This is (-12,48,36), (-60,48,36), (-12,132,36) and (-60,132,36) in inches.

Now, a volume of the rectangular prism is given as 252 square feet. We know from the first two given sets of coordinates that the absolute width of the prism is 4 (change from -1 to -5 when comparing the first two coordinate sets), and the second dimension is found by comparing the first and third coordinate sets (change from 4 to 11), so the value of the second dimension is 7. Note that we do not compare the first and fourth, or second and third coordinate sets, because these comparisons show changes for two numbers, which means a diagonal (it is not a dimension that we need in order to find the third dimension plugging in the known volume). So, the third dimension is $\frac{252}{(7)(4)} = 9$ feet.

Now, our coordinates show as (*x*,*y*,*z*) in which *z* is a third dimension that we found (height of the prism in this case). In inches, this height is 9(12) = 108 inches. So, all we need to do now is to simply add 108 to the existing *z*-value of 36 to each of the known coordinate sets, and this will give us the desired top four corner edges. This gives us (-12,48,144), (-60,48,144), (-12,132,144) and (-60,132,144). Note that this prism lies in the second octant (where all *x* values are negative, and *y* and *z* are positive).

Answer: B

64. The following figure shows a rectangle short of three small cut out rectangles, of which the bottom two are identical. If the dimensions of the parent rectangle are 24 and 38, find the product of the diagonals shown.

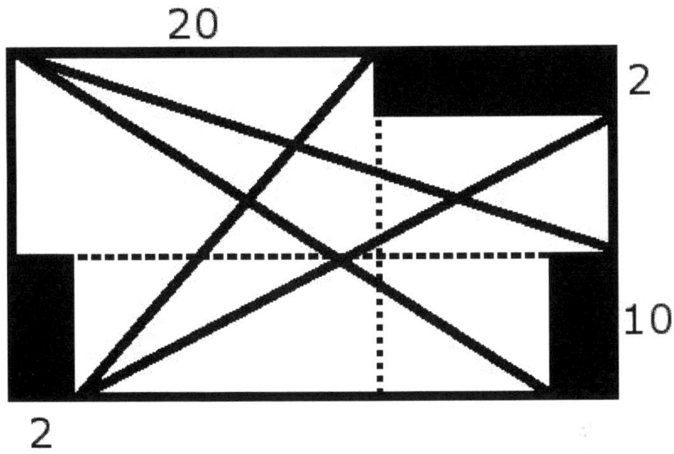

A. 2,362,500

B. 2,461,273

C. 3,322,466

D. 2,217,717

This problem simply requires one to find all shown diagonals correctly and then find their product. We will first find the two diagonals with positive slope. The base of the triangle made up by the longer diagonal of the two is 38 – 2 = 36. The height is 24 – 2 = 22. The diagonal is then 42.19004622. We used the *P. T.* to find it.

Now, the base for the steeper diagonal is 20 – 2 = 18. Height is 24. Diagonal is then 30. Now we will find the diagonals with negative slope. Base for the longer diagonal is 38 – 2 = 36 (knowing that the bottom two small rectangles are identical). Height is 24. Diagonal is then 43.26661531. The last diagonal in question has a base that is 38, and height 24 – 10 = 14. Diagonal is 40.49691346. Thus, the product of the four diagonals that we found is about 2,217,717.

Answer: D

65. The larger of the two similar right triangles has an altitude that is one of its sides, and measures 5. If the area and base of the smaller triangle is 68 and 13, find the height of the larger triangle that is perpendicular to its hypotenuse.

A. 4

B. 5

C. 6

D. 8

Here we use the concept of proportion to find the missing side (name it base *b*) for the large triangle. Since the area of the small triangle is 68, the product of its base and height *h* is 136 (area is base times height divided by 2). So, 13 • *h* = 136, making *h* = 10.46153846. Now, the proportion will be as follows: $(\frac{h}{13}) = (\frac{5}{b})$, and we solved for *h* already, so the large triangle base *b* is 6.213235295.

Now, we need to figure out a different height than what was given (the new height is perpendicular to the hypotenuse). To find this height, we must first find the hypotenuse: we use the *P. T.* and known base and height to find it, and we get 7.975229955 for the hypotenuse. If we rotate the entire large triangle so that the hypotenuse becomes the base, we can use the area formula to find the required height: we know the area, because we solved for original height and base, so our area is $\frac{6.213235295(5)}{2}$ = 15.53308824. Now we can solve for the required new height knowing the area and new base (rotated hypotenuse): 15.53308824 = $\frac{7.975229955(new\ height)}{2}$, giving us the new height as 3.895333007.

Answer: A

66. The volume of the rectangular prism is 866. A similar smaller rectangular prism has the longest diagonal measuring 7. The ratio of sides of the larger prism to the smaller prism is 3:1, and the length of the larger prism is 11. Find the dimensions of the smaller rectangle.

A. 11, 2, 4

B. 11, 7, 4

C. $\frac{11}{3}$, 6, 2

D. $\frac{11}{3}$, 2, 4

In this example we use the fact that the ratio of volume of similar figures is the cube of the side ratio of the same similar figures. Since the volume of the larger prism is 866, volume of the smaller prism must be $\frac{866}{27}$ or 32.07407407 (27:1 is the cube of the side ratio that is 3:1). Since the length of the larger prism is 11, length of the smaller prism is $\frac{11}{3}$ (side ratio is 3:1). Now, since 7 is the longest diagonal in the smaller prism, we know that $7 = \sqrt{l^2 + w^2 + h^2} = \sqrt{(\frac{11}{3})^2 + w^2 + h^2}$, meaning that $49 = (\frac{11}{3})^2 + w^2 + h^2$. Since $32.07407407 = (\frac{11}{3})wh$, $wh = 8.74747474$, and $h = \frac{8.74747474}{w}$. We can plug this in the equation $49 = (\frac{11}{3})^2 + w^2 + h^2$, so that it becomes $49 = (\frac{11}{3})^2 + w^2 + (\frac{8.74747474}{w})^2$.

We can now solve for width w. Simplifying and rewriting this expression, we get $w^4 - 35.55555556w^2 + 76.51831445 = 0$. Using the quadratic formula, we get $w^2 = 33.25456906$, or $w = 5.766677471$ (we could have used the second smaller root; it would not have changed our final answer). This makes $h = 1.516900293$.

Answer: C

67. The longest diagonal of an isosceles triangular prism is 12.5, and slanted height 5. The area of the largest face of the prism is 66 (with length the longest side of this face). If the ratio of sides of this prism and a similar larger prism is 1:5, find the volume and surface area of the larger prism.

A. V = 4,989, S_A = 16,458

B. V = 24,093, S_A = 3,147

C. V = 3,147, S_A = 24,093

D. V = 16,458, S_A = 4,989

This problem is similar to the previous problem, and requires the use of proportions as well as diagonal equations. Since the area of the largest face is 66, this means that it will have the longest diagonal associated with the dimensions making this largest face. So, $12.5^2 = b^2 + l^2$, and $lb = 66$. If we set $l = \frac{66}{b}$, we can use this in the first equation involving the diagonal, namely as $b^2 + (\frac{66}{b})^2 = 156.25$. Simplifying and rewriting this equation, we get $b^4 - 156.25b^2 + 4{,}356 = 0$. Solving for b using the quadratic formula, we get $b^2 = 36.32170318$, or $b = 6.026748973$. This is the base for the smaller prism (we could not have used the second root, because it is larger – in fact it will be our length, and length is the largest dimension since it is part of the largest face of the prism). Length l is then $\frac{66}{b} = 10.95117787$. To find height h, we divide the base (6.026748973) in half (since the prism has isosceles triangle faces), and use the *P. T.* also knowing the given slanted height 5. We get $h = 3.98993411$. Volume is then $\frac{lbh}{2} =$ 131.6678256.

Since the side ratio of the larger prism to smaller prism is 5:1, the volume ratio will be the cube of that – 125:1. So, the volume of the larger prism is 125(131.6678256) ≈ 16,458. To find surface area, we first convert the dimensions we found for the smaller prism to larger prism knowing the side ratio 5:1. Length is then 54.75588937, slant height = 25, base = 30.13374487, height = 19.94967055. We also need to convert 66 (largest face of the smaller prism) to larger prism, knowing that the area ratio is the square of the side ratio 5:1, so it is 25:1, and 66 turns to 66(25) = 1,650. Surface area of the larger prism is then $2(\frac{bh}{2})$ + 2(slant height • length) + 1,650 ≈ 4,989.

Answer: D

68. If the radius of a circle is 40*y* meters and its area is 5π square meters, find the area (in square meters) of the circle whose radius is 8*y* kilometers.

A. 200,000π m^2

B. 200π m^2

C. 40,000π m^2

D. 1,000π m^2

In this problem we use the concept of proportion of radius and volume for two circles that are compared. The ratio of radiuses is (40*y* m):(8*y* km), or (40*y* m):(8,000*y* m), which means the simplified ratio is 1:200. This means that the area ratio will be the square of that, or 1:40,000. Since the area of the first circle is 5π square meters, we simply multiply this amount by 40,000 to get the area for the second circle, or 200,000π square meters.

Answer: A

69. In the following right triangle *ABC*, measure of angle *A* is 20 degrees, and side *AB* is 10. Segment *DB* is perpendicular to hypotenuse *CA*. What is the value of segment *DB*?

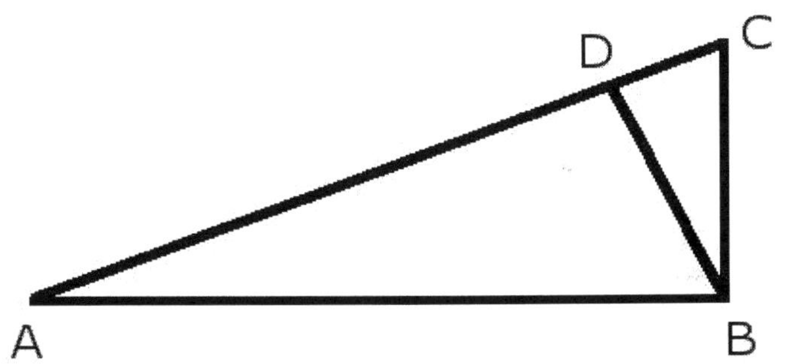

A. 5

B. 4

C. 2

D. 3

In this example, we use the concept of trigonometry to find segment DB. Since angle A measures 20 degrees, and segment AB is 10, we will have $\tan(20^o) = \frac{BC}{10}$, making BC =

3.639702343. Angle C measures 180 – 20 – 90 = 70 degrees. Now will use this as angle of reference: $\sin(70^o) = \frac{DB}{3.639702343}$, making DB = 3.420201434.

Answer: D

70. In a rectangular prism, the longest diagonal makes a 25 degree angle of elevation. If the height of the prism is 9 feet, and the length is 3 times the width, find the surface area (in square inches) and volume (in cubic inches) of the prism.

A. V = 1,006 in ^{3}, S_A = 67 in ^{2}

B. V = 9,623 in ^{3}, S_A = 1,737,991 in ^{2}

C. V = 67 in ^{3}, S_A = 1,006 in ^{2}

D. V = 1,737,991 in ^{3}, S_A = 9,623 in ^{2}

We first find the diagonal, using a right triangle with height h = 9 and making our diagonal a hypotenuse in this triangle. Since the angle between the hypotenuse and base face measures 25 degrees and the opposite side to the angle is the height h, *we* will have $\sin(25^o) = \frac{9}{hyp}$, making our hypotenuse 21.29581425.

Now, we know that a 3-dimensional diagonal is expressed by the square root of the sum of three squared dimensions. It is given that the length is three times the width. So, if we let width be x, length will be 3x, and the diagonal (hypotenuse) will be 21.29581425 = $\sqrt{81 + x^2 + 9x^2}$, so that 453.5117045 = 81 + 10x^2. This makes x (the width) equal 6.103373694, and length 18.31012108. Volume is then whl = 1,005.781602, and in cubic inches this is (1728)(1,005.781602) ≈ 1,737,991. Surface area is 2l + 2w + 2h = 66.82698955, and in square inches this is 144(66.82698955) ≈ 9,623.

Answer: D

71. A sphere is tangent to the centers of the faces of the cube that perfectly encloses it. If the longest diagonal of the cube is 28 meters, find the volume (in cubic kilometers) and surface area (in square kilometers) of the sphere.

A. V = 2.2 • 10^{-6} km ^{3}, S_A = 8.2 • 10^{-4} km ^{2}

B. V = 8.2 • 10^{-4} km ^{3}, S_A = 2.2 • 10^{-6} km ^{2}

C. V = 821 km ^{3}, S_A = 2,212 km ^{2}

D. V = 2,212 km ^{3}, S_A = 821 km ^{2}

Since the cube dimensions are all equal, we can name each as *x*. The longest diagonal of the cube will be 28 = $\sqrt{3x^2}$, and *x* = 16.16580754. The radius of the sphere is half of *x*, or 8.08290377. Volume of the sphere is then $\frac{4\pi r^3}{3}$ = 2,212.029092 cubic meters. Converting to cubic kilometers (where 1 cubic km = 1,000,000,000 cubic m), we get 2.2 • 10^{-6} cubic kilometers as our volume. Surface area of the sphere is $4\pi r^2$ = 821.0028804 square meters. Converting to square kilometers (1 square km = 1,000,000 square m), we get 8.2 • 10^{-4} square km.

Answer: B

72. The following figure shows two mutually perpendicular illustrations of an ellipsoid that is inscribed in a sphere. The left view shows a triangle leg that matches the sphere's radius in length. If *a* = 15 feet, m<*A* = 33°, m<*B* = 72°, find the volume (in square yards) of the sphere region outside the ellipsoid.

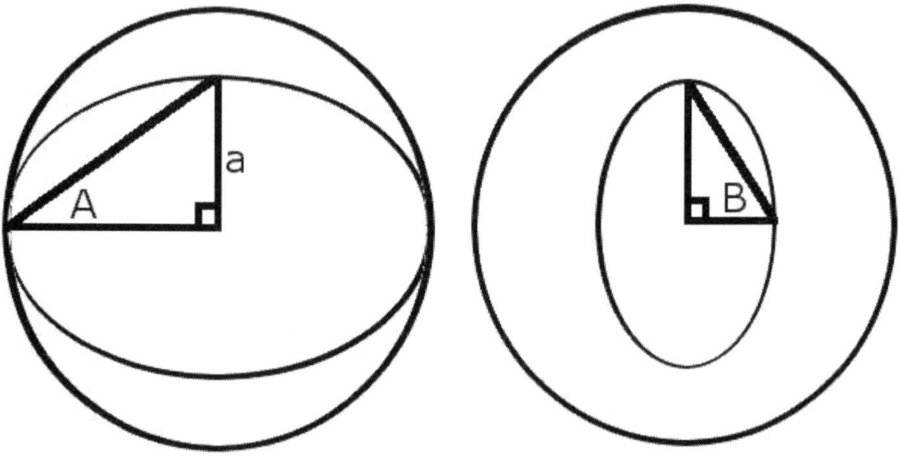

A. 4,950 yd ^{2}

B. 44,546 yd ^{2}

C. 7,073 yd ^{2}

D. 786 yd ^{2}

If you turn the sphere about the vertical imaginary axis, you will notice that radius *a* (triangle height *a*) does not change. So, side *a* is also shown in the right image as a triangle height. Since $a = 15$, adjacent triangle side (principal radius r_1 of the ellipsoid) in the left image is found: $\tan(33^{\circ}) = \frac{15}{r_1}$, giving $r_1 = 23.09797446$. Second principal radius is found similarly: $\tan(72^{\circ}) = \frac{15}{r_2}$, giving $r_2 = 4.873795444$.

The third principal radius of the ellipsoid is the height we already know, 15. Sphere volume is calculated using the largest principal radius of the ellipsoid that matches the sphere radius (left image), namely 23.09797446. Sphere volume is then $\frac{4\pi r^3}{3} =$ 51,619.08473. Ellipsoid volume is calculated in the same way as the sphere, only that the three principal radiuses are distinct: $\frac{4\pi(15)(23.09797446)(4.873795444)}{3}$ = 7,073.283462. Thus, the area of the sphere not including the ellipsoid area is 51,619.08473 – 7,073.283462 = 44,545.80127. Converting square feet to square yards (1 sq yd = 9 sq ft), we get volume as about 4,950.

Answer: A

73. A gigantic car wheel makes 77 revolutions per 12 minutes. What is the measure of central angle *x* illustrated in a car wheel circle below given that the speed of the car is 297 miles per hour, and the longest side of the triangle is 972 feet long?

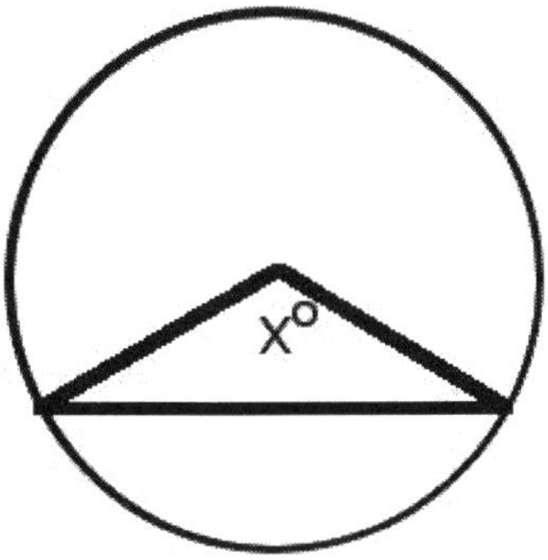

A. 94 degrees

B. 101 degrees

C. 108 degrees

D. 97 degrees

In this problem, we must match the whole distance traveled with the circumference of the wheel (which is one wheel revolution) together with the fact that there are 77 wheel revolutions per 12 minutes, or 77(5) = 385 revolutions per hour. We set 297 = 77(5)(2π*r*), and get *r* = 0.12277667 miles.

Now, the longest side (base) of the triangle can be divided in half (and the triangle itself to small right triangle) because the triangle is isosceles, so that 972 turns into 486. However, this is in feet, so we this distance to miles (1 mile = 5,280 feet), so that 486 feet is 0.092045455 in miles. Now, our small triangle hypotenuse is the radius that we found, and the base is 0.092045455. The angle in question has this base as the opposite side, so we find it using the $\sin^{-1}$ (arcsine) concept. We find the angle sine of which produces the quotient $\frac{0.092045455}{0.12277667}$. So, using arcsine we have

$\sin^{-1}(\frac{0.092045455}{0.12277667})$ = 48.56424639 degrees. We must now remember to multiply this angle measure by 2, because we had divided the original large isosceles triangle in half, so we get the angle measure to be about 97 degrees.

Answer: D

74. The following figure shows an equilateral polygon. If the line drawn from angle *B* is perpendicular to side *AC* and measures 11 feet, find the area (in yd^2) of the triangle *ABC* (not shown).

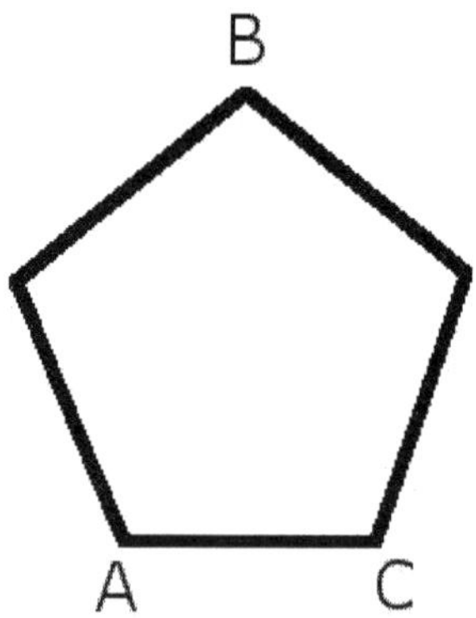

A. 2 square yards

B. 20 square yards

C. 39 square yards

D. 4 square yards

The perpendicular side to AC from angle B makes a triangle if we add side AB or AC to it, and these two right triangles will make a large triangle ABC whose area we are required to find. It is sufficient to find the area of one of these two triangles and then multiply it by 2 (because the large triangle ABC is isosceles since the pentagon is equilateral). Perpendicular line from angle B to AC will divide AC in half (half of base for the large triangle ABC).

Now, we do not have any dimensions given for the pentagon, so we must go with trigonometry and use the angle B for any one of the two right triangles we created. Note that line AC makes an isosceles triangle that has base AC. Now, since each angle for the pentagon must measure $\frac{180(5-2)}{5}$ = 108 degrees, each of the two equal angles in this isosceles triangle measures $\frac{180-108}{2}$ = 36 degrees. This makes angle B for the large triangle ABC in this problem measure 108 – 2(36) = 36 degrees (we multiplied 36 by 2 in the subtraction because line BC also creates the same isosceles triangle with the same angles). Dividing this new angle B measure by two (because we are working with one of the two small right triangles that is half of the large triangle ABC), we get 18 degrees.

Now we have a right triangle with angle measuring 18 degrees and adjacent side that comes from angle B and is perpendicular to side AC. We now need the opposite side (half of base for the large triangle ABC), so we have tan(18°) = $\frac{opp}{11}$, and the opposite side is 3.574116659. To get the base for the large triangle ABC, we multiply the result by 2 to get base *B* = 7.148233317. Height of the triangle ABC is 11 (given by the problem). Area of triangle ABC is then $\frac{11(7.148233317)}{2}$ = 39.31528324. This is in square feet. Converting to yards (1 square yard = 9 square feet), we get 4.368364804.

Answer: D

75. In the following figure, if the height is 6 feet, one of the bases 15 feet, and area 78 square feet, find the measure of angle *x*.

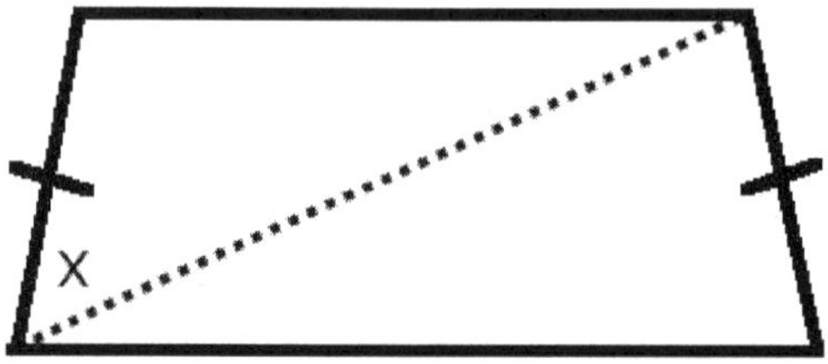

A. 54°

B. 62°

C. 47°

D. 59°

This problem can be done in two ways: one way is to use the *Law of Cosines* after finding the trapezoid second base, slanted height (this is one of the identical sides of a trapezoid) and diagonal to find angle *x*, and the other way is to subtract angle other than *x* (this is the angle between the large hypotenuse (diagonal) and longest base of the trapezoid) from the second largest angle in the triangle (where the slant height is the hypotenuse and given vertical height 6) to get measure of angle *x*. Let us try the other way first. See the figure below for more clarity.

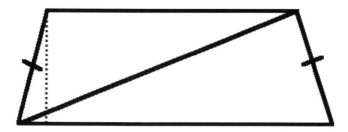

First, we need to find the other trapezoid base *y* using the area given and vertical height. We have $78 = (15 + y)(\frac{6}{2})$, so that *y* (second trapezoid base) is 11. To find the diagonal, we need to understand the horizontal distance it covers: since the two slant heights of the trapezoid are equal, the respective horizontal distances that these slant heights cover are also equal – which means that if the longer base is 4 feet longer than the shorter base, each slant height takes 2 feet of horizontal distance, so the diagonal takes 15 – 2 = 13 feet of horizontal distance. Diagonal is then $\sqrt{13^2 + 6^2}$ = 14.31782106.

The angle measure between this diagonal (hypotenuse in this triangle) and large base is then calculated using arcsine: $\sin^{-1}(\frac{6}{14.31782106})$ = 24.77514057 degrees. Now we will find the angle measure between the slant height and the same base (2 feet) because we need to subtract the angle measure we found from this angle to find angle *x*, so we use arctangent: $\tan^{-1}(\frac{vertical\ height}{base}) = \tan^{-1}(\frac{6}{2})$ = 71.56505118 degrees. Thus, angle *x* measures 71.56505118 – 24.77514057 = 46.78991061 degrees.

Now, let us solve this using the first way: if we divide this triangle using the provided diagonal, angle *x* can be calculated using the *Law of Cosines* using slant height, diagonal and top trapezoid base: slant height is $\sqrt{2^2 + 6^2}$ = 6.32455532. Thus, we can set this up using the law: $11^2 = 6.32455532^2 + 14.31782106^2 - 2(6.32455532)(14.31782106) \cdot \cos(x)$, and using algebra and arccosine ($\cos^{-1}$) we get $x \approx 47$ degrees.

Answer: C

76. In the following figure, an equilateral triangle perfectly inscribes a circle. If the height of the triangle is 14 feet, find the perimeter and area of the region marked *x* (using inches).

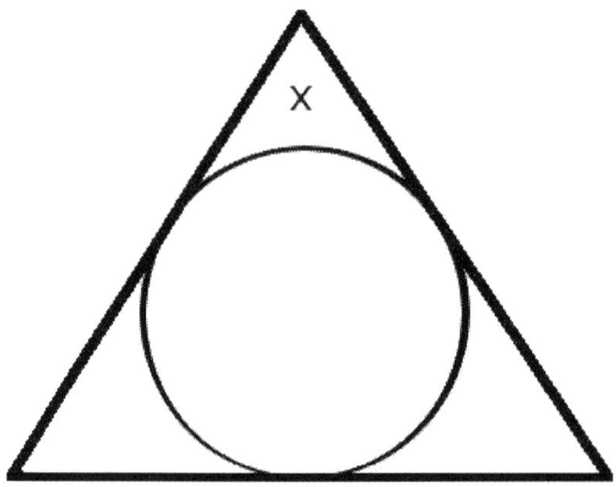

A. P = 15 in, A = 26 in ^{2}

B. P = 2,148 in, A = 311 in ^{2}

C. P = 26 in, A = 15 in ^{2}

D. P = 311 in, A = 2,148 in ^{2}

We need to find the radius of the circle as well as the side of the equilateral triangle to work with perimeter and area. See the following figure for more detail. We know that the height given (14) can be any one of the three heights, because each side is equal in this triangle, and so are its heights. We can form a triangle using the dashed lines as seen in the figure above, with hypotenuse being the height length (14). Now, we know that each angle in the equilateral triangle measures 60 degrees, so the height divides the triangle in half, so the triangle we created has the smallest angle that measures 30 degrees. Vertical dashed side of the triangle is then: $\sin(30^{o}) \cdot 14 = 7$.

Next, we can find the radius of the circle using the concept of direct proportion, shortening our hypotenuse for the second imaginary triangle where the hypotenuse will stop at the center of the circle, and vertical triangle side will be the circle radius. This will make a proportion for the two imaginary triangles as follows: $\frac{14}{7} = \frac{14-r}{r}$, solving for radius r we get $\frac{14}{3}$.

Now, we find the side of the original equilateral triangle, working with the half of the equilateral triangle created by the long dashed hypotenuse: cos(30°) = $\frac{14}{side}$, making *side* equal 16.16580754. Area of the equilateral triangle is then $\frac{16.16580754^2\sqrt{3}}{4}$ = 113.1606528, and area of the circle is $\pi(\frac{14}{3})^2$ = 68.41690668. Area marked *x* is then $\frac{113.1606528 - 68.41690668}{3}$ = 14.91458204 square feet. In square inches, this is 14.91458204(144) ≈ 2,148. The perimeter of the region marked *x* consists of the sum of 2 halves of the triangle side (basically one full side) added to one-third of the circle circumference: 16.16580754 + $\frac{2\pi(\frac{14}{3})}{3}$ = 25.93965135. In inches, this is about 311.

Answer: D

77. In the following figure, an isosceles triangle shows a hypotenuse *AC* that is 21 meters long. Segment *BD* is perpendicular to *AC*, and *BE* = *ED*. Find segment *EF* expressed in yards (nearest tenth).

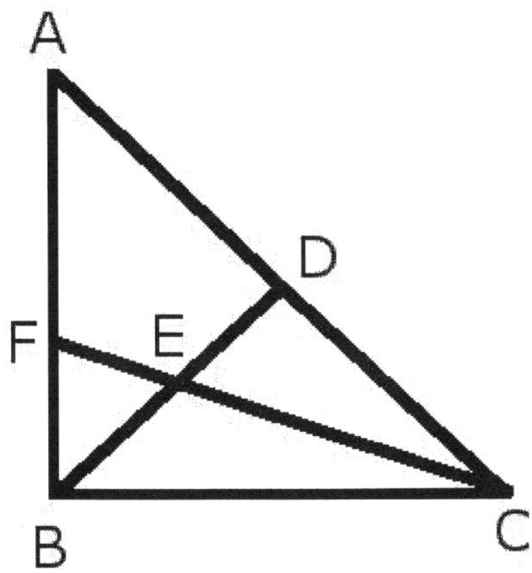

A. 1.1 yd

B. 12.8 yd

C. 4.3 yd

D. 3.9 yd

First, we notice that since BD is perpendicular to AC, it divides AC in half and also creates two isosceles triangles (since triangle ABC is isosceles). This means that DC = BD = $\frac{21}{2}$ = 10.5. Since BE = ED, each of them will be $\frac{10.5}{2}$ = 5.25.

We can now work with right triangle EDC to find segment EC which is the hypotenuse of this triangle: EC is then $\sqrt{5.25^2 + 10.5^2}$ = 11.73935688. We must also find segment FC in order to find the segment EF. We will use the triangle FBC to do that. However, we only know one side in this triangle, the base (also the base of triangle ABC) that is calculated using the *P. T.* from the given triangle ABC hypotenuse of 21. This base is calculated as follows: since the two sides of triangle ABC are the same, we will have $21^2 = x^2 + x^2 = 2x^2$, therefore the base *x* is 14.8492424. Again, we cannot find FC using the *P. T.* for triangle FBC as we did with triangle EDC, because we do not know segment FB. So we must use the angle BCF using trigonometry. To find the measure of this angle, we must first find measure of angle DCE and then subtract that from 45 degrees (measure of angle BCD) to find measure of angle BCF.

We can easily find measure of angle DCE, since we know segments ED and DC. We use the arctangent to find the angle measure. Angle DCE measure is then $\tan^{-1}(\frac{5.25}{10.5})$ = 26.56505118 degrees. Angle BCF measure is then 45 – 26.56505118 = 18.43494882 degrees. Segment FC is then calculated: $\cos(18.43494882°) = \frac{14.8492424}{FC}$, making FC = 15.65247584. Thus, segment EF = FC – EC = 15.65247584 – 11.73935688 = 3.91311896 m. In yards (where 1 m = 1.093613298 yd) this is about 4.3.

Answer: C

78. The following figure shows a conglomerate containing a part of a circle, rectangle (its width equals the radius of the circle) and isosceles triangle. The dashed diagonal is 23 feet long. Find the perimeter (in inches) and the area (in square yards) of the figure.

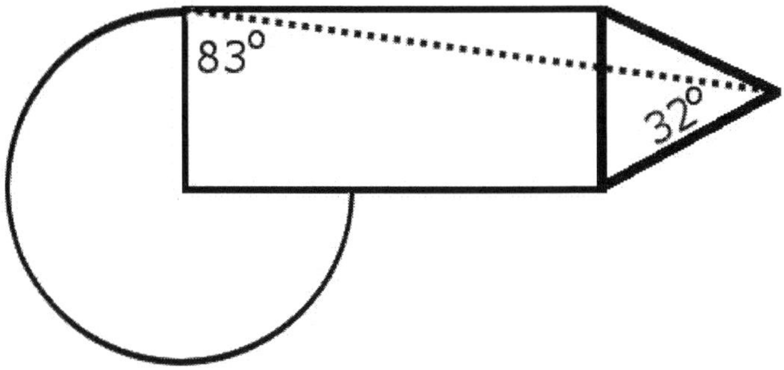

A. P = 813 in, A = 21 yd ^{2}

B. P = 21 in, A = 813 yd ^{2}

C. P = 203 in, A = 26,665 yd ^{2}

D. P = 813 in, A = 26,665 yd ^{2}

To find the perimeter and area of the figure below, we need to find the width of the rectangle (this is also the radius of the circle), length of the rectangle, and two equal slanted heights of the isosceles triangle.

First, we notice that we can make a right triangle using the dashed line as the hypotenuse (given as 23) of the triangle. This will give us half of the width (label it *w*) of the rectangle. We have $\cos(83^{\circ}) = \frac{w}{23}$, making *w* = 2.802994898. This makes the width *W* of the rectangle 2(2.802994898) = 5.605989797. We now also found the radius of the circle that is also 5.605989797. The horizontal distance *d* that the hypotenuse of 23 covers is: $\sin(83^{\circ}) = \frac{d}{23}$, making *d* = 22.82856149. Since the angle measure of 32° was given in the original figure in the question, and this angle is between the dashed hypotenuse and isosceles triangle slant height, and knowing that the angle measure in this figure between the dashed hypotenuse and vertical height is 7 degrees (why?), we know that the angle between the slant height of the isosceles triangle and its vertical height measures 32 – 7 = 25 degrees.

Since the vertical height of the triangle divides the base (rectangle width) in half (and creates two small right triangles), we can find this vertical height *h* because we know the smallest angle measure and half of the width of the rectangle (base of the small triangle): $\tan(25^o) = \frac{2.802994898}{h}$, making vertical height h = 6.011041957. Hypotenuse of the small triangle (one of the slant heights of the large triangle) is then $\sqrt{6.011041957^2 + 2.802994898^2}$ = 6.632450965. We could have used trigonometry to find this as well. Note that we can now find the length l of the rectangle by subtracting the vertical height from the horizontal distance that the dashed hypotenuse covered: 22.82856149 – 6.011041957 = 16.81751953. The perimeter P = $(\frac{3}{4}$ of the circle circumference + sum of two slant heights of the triangle + rectangle length + (rectangle length – width)) = $(\frac{3}{4})(2\pi(5.605989797))$ + 2(6.632450965) + 16.81751953 + (16.81751953 – 5.605989797) = 67.71155574 feet. In inches this is about 813 inches. The area of the figure is then $(\frac{3}{4}$ the area of the circle + area of the rectangle + area of the triangle (this is a rectangle width times the triangle vertical height divided by 2)) = $(\frac{3}{4})(5.605989797^2\pi)$ + 5.605989797 • 16.81751953 + $\frac{5.605989797(6.011041957)}{2}$ = 185.1761736 square feet. In square yards, this is about 21.

Answer: A

79. If the square side is 12, which expression is valid regarding the angles *x*, *y* and *z*?

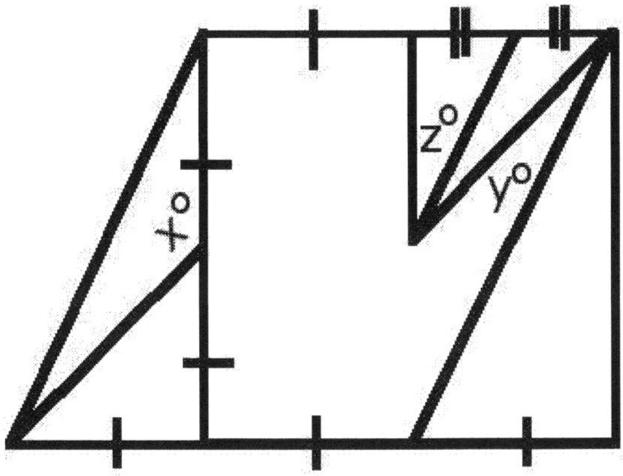

A. $(_{m<}x - 180) = (_{m<}y + _{m<}z)$

B. $(_{m<}z - _{m<}y) < 0$

C. $(_{m<}x + _{m<}y) = (_{m<}z + 128)$

D. $(2_{m<}z + 3_{m<}y) \approx (\frac{4}{5})_{m<}x$

To find measure of angle *x*, we will subtract the adjacent angle measure of *x* (angle that is supplementary with *x*) from 180 degrees.

The adjacent angle measures 45 degrees, since the adjacent and opposite sides of this angle are equal. Thus, angle *x* measures 180 – 45 = 135 degrees. The opposite side of angle *z* is $\frac{1}{2}$ of the half of the square side, that is, 3. The adjacent side of angle *z* is half the square side, namely 6. Thus, angle *z* measures $\tan^{-1}(\frac{3}{6})$ = 26.56505118 degrees. Angle *y* can be calculated by subtracting the adjacent angle measure (angle in the large right triangle whose one side is the square side) from 45 degrees. The adjacent angle to *y* belongs in a large right triangle whose sides are 12 and 6. Thus, the adjacent angle to *y* is equal in measure to angle *z* (why?). Thus, angle *y* measure is then 45 – 26.56505118 = 18.43494882 degrees. Plugging in the known values into each choice provided, we can see that the last choice is correct.

Answer: D

80. A right circular cone's base circumference is 19π meters. If the angle between the base and slant height of the cone measures 52 degrees, find the volume (in cubic decimeters) and surface area (in square kilometers) of the cone.

A. V = 1,149,184 dm^3, S_A = ($7.44 \cdot 10^{-6}$) km^2

B. V = ($7.44 \cdot 10^{-6}$) dm^3, S_A = 1,149,184 km^2

C. V = 1,149 dm^3, S_A = 744 km^2

D. V = 744 dm^3, S_A = 1,149 km^2

Since the base circumference of the cone is 19π, the radius of the base circle is 9.5. We can think of a cone as an isosceles triangle, with radius being half of the base of the triangle. We will then have two equal right triangles with base 9.5.

Using trigonometry, we find the vertical height *h* knowing the angle of 52 degrees: $\tan(52^o) = \frac{h}{9.5}$, giving height *h* = 12.15944551. Volume of the cone is then $\pi(9.5^2)(\frac{12.15944551}{3})$ = 1149.184076 cubic meters. In cubic decimeters this is about 1,149,184. Surface area is $9.5\pi(9.5 + \sqrt{12.15944551^2 + 9.5^2})$ = 744.0557448 square meters. In square kilometers (1 square kilometer equals 100,000,000 square decimeters) this is $7.44 \cdot 10^{-6}$.

Answer: A

81. The following figure shows an isosceles triangular prism. If one of the identical sides of the isosceles face is 7 feet long, find the measure of the angle between the left slanted height of the triangle face and the dashed hypotenuse, the volume (in cubic inches) and surface area (in square miles) of the prism.

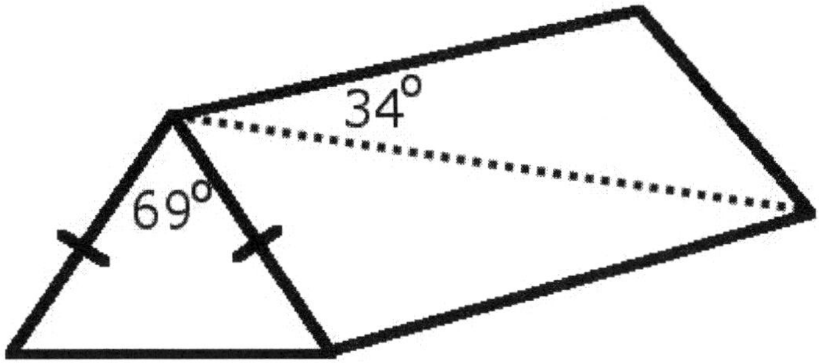

A. 78°, V = 410,178 in ^{3}, S_A = $(9.8 \cdot 10^{-6})$ mi ^{2}

B. 80°, V = 273 in ^{3}, S_A = 237 mi ^{2}

C. 75°, V = $(9.8 \cdot 10^{-6})$ in ^{3}, S_A = 410,178 mi ^{2}

D. 81°, V = 237 in ^{3}, S_A = 273 mi ^{2}

We will first find the volume and surface area of the prism. Since the triangle face is isosceles, we can divide the face in half, thus creating two small right triangles. The top angle measure becomes 34.5 degrees. Since the slant height (hypotenuse of one of the small right triangles created) is 7, we can find vertical height h: $\cos(34.5^\circ) = \frac{h}{7}$, making $h = 5.76888332$. Half of the base of the large triangle face is then: $\sin(34.5^\circ) = \frac{b}{7}$, making $b = 3.964843658$, and thus the base B of the large triangle (base of the triangle face) is 2(3.964843658) = 7.929687317.

Now, the length l is found the same way using the second given angle: $\tan(34^\circ) = \frac{7}{l}$, making length $l = 10.37792678$. Volume of the prism is then $(7.929687317)(5.76888332)(\frac{10.37792678}{2}) = 237.3714181$ cubic feet. In cubic inches this is $1728(237.3714181) \approx 410{,}178$. The surface area is $2(7.929687317)(\frac{5.76888332}{2}) + 2(7)(10.37792678) + 7.929687317(10.37792678) =$

273.3301302 square feet. In square miles (1 square mile = 27,878,400 square feet) this is $\left(\frac{273.3301302}{27,878,400}\right) \approx 9.8 \cdot 10^{-6}$.

Now, to find the angle measure between the dashed hypotenuse and the left slant height of the triangle face of the prism, it is best to use the *Law of Cosines*. We will create a triangle consisting of the hypotenuse of the base face of the prism, slant height of the prism and the dashed hypotenuse shown in the figure. We first find the dashed hypotenuse: $\sqrt{7^2 + 10.37792678^2}$ = 12.51804155. We now only need to know the hypotenuse of the base prism face (to be calculated by the dimensions of length and base of the triangle face). The hypotenuse is then $\sqrt{7.929687317^2 + 10.37792678^2}$ = 13.06067782. The angle *A* whose measure we need to find has the hypotenuse we just found as its opposite side, so it must be isolated on one side of the law equation: $13.06067782^2 = 7^2 + 12.51804155^2 - 2(7)(12.51804155) \cdot \cos(A)$, and rewriting and simplifying the equation we get $A = \cos^{-1}(0.200396815) \approx 78$ degrees.

Answer: A

82. The following figure shows two mutually perpendicular shots of a rectangular prism enclosing an ellipsoid that is tangent to the centers of each dimension of the rectangular prism. Both images include the length of the prism, which is 15 meters. The width is the smallest dimension of the prism. Central exterior angles are also shown. Find the volume of the prism (in cubic feet) outside the volume of the ellipsoid.

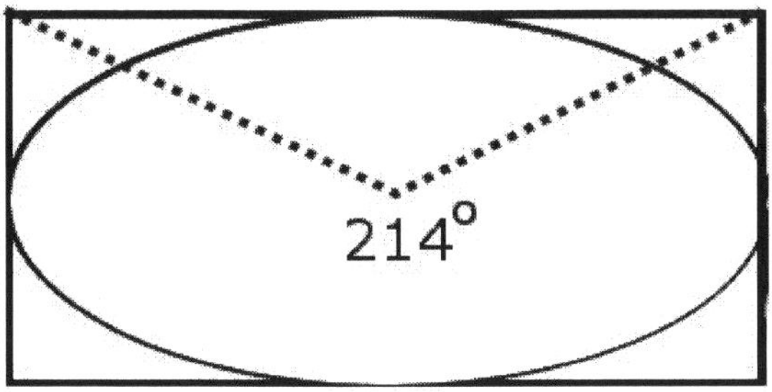

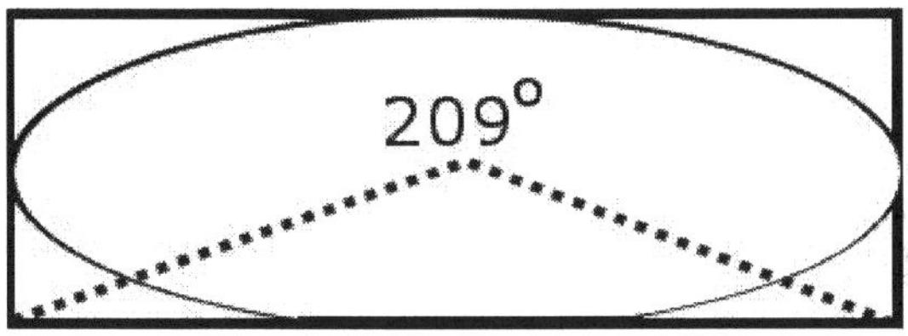

A. 127 ft^3

B. 267 ft^3

C. 9,424 ft^3

D. 4,490 ft^3

Since the length is included in both images, the left image contains length and height, and the second image contains length and width (width is the shortest side according to the problem). We will first work with the first image to find the height: we will work with the isosceles triangle (half of it to be precise) bounded by the dashed lines and length of the base. The interior angle measure (largest angle of the triangle we created) must add up to 360 with the given 214 angle measure, so the interior angle measures 360 – 214 = 146 degrees. Since the triangle that we created is isosceles, we divide it in half (and the angle of 146), making two small right triangles each with angle measure 146/2 = 73 degrees.

The dimensions of any one of these two right triangles include half of the height and half of the length. Since we know that half of the length is $\frac{15}{2}$ = 7.5, we use trigonometry to find half of the height: tan(73°) = $\frac{7.5}{\frac{h}{2}}$, making height half equal 2.292980111 and height h = 4.585960222. Note that all three halves of our three dimensions (length, height, width) are also the principal *radiuses* of the enclosed ellipsoid (we will need them to calculate the ellipsoid volume).

Next, we find the width: working now with the second image, we take a look at the isosceles triangle visible by the bounds of the dashed lines and the length. The interior angle measure of this triangle must add up to 360 with the 209 angle given, so it must be 360 – 209 = 151 degrees. Dividing this in half, we have an angle measure of 75.5 in the small right triangle created. Since the length half is 7.5 (dimension of the right triangle), the width half ($\frac{w}{2}$) is then: $\tan(75.5^\circ) = \frac{7.5}{\frac{w}{2}}$, making $\frac{w}{2}$ = 1.939631883 and w = 3.879263765. Prism volume is then (3.879263765)(15)(4.585960222) = 266.8522398 cubic meters. Ellipsoid volume is $(\frac{4\pi}{3})(7.5)(2.292980111)(1.939631883)$ = 139.723506 cubic meters. The resulting volume in the prism not occupied by the ellipsoid is 266.8522398 – 139.723506 = 127.1287338 cubic meters. Translating to cubic feet (1 cubic meter = 35.31466672 cubic feet), our answer turns out to be about 4,490.

Answer: D

83. In the following figure, a right triangular pyramid (with identical 3 triangle faces, and base an equilateral triangle) is shown. If the perimeter of the base face is 33 feet, and one of the longest slanted heights is 11 feet, find the surface area (in square inches) and volume (in yd ^{3}) of the pyramid.

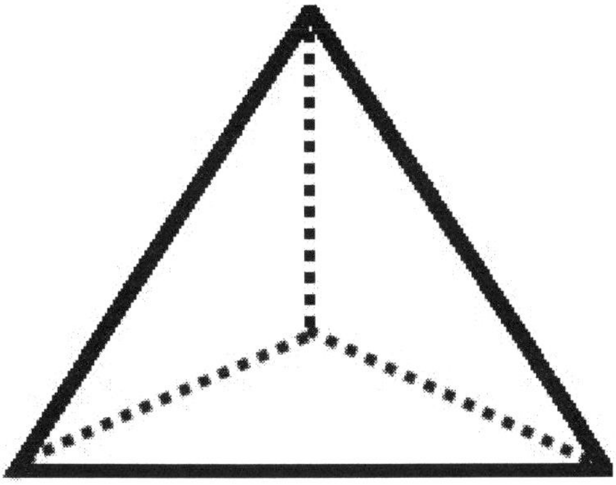

A. $S_A = 210\ in^2$, $V = 271{,}054\ yd^3$

B. $S_A = 30{,}179\ in^2$, $V = 157\ yd^3$

C. $S_A = 30{,}179\ in^2$, $V = 6\ yd^3$

D. $S_A = 210\ in^2$, $V = 6\ yd^3$

Since the base face perimeter is 33 feet (and base is an equilateral triangle), each side of the base is 11 feet long. Since the pyramid has 3 identical triangular faces, their apex (tip) must be connected by a straight and perpendicular line going through the center of the base face. The longest slant height is the shared border line (edge) between each slant pyramid face, and is given as 11 feet. Since the 3 triangular faces are identical, there are 3 such border lines (longest slanted heights that equal 11 feet). This makes all 4 faces of the pyramid identical. To find the volume, we need to find the vertical height that connects the center point of the base face with the apex of the pyramid (tip), and find the height of the base face. Surface area will be easy to find, because all four faces are identical. For the volume, to find the vertical height (to the tip) we need to first understand the base face: see the figure below.

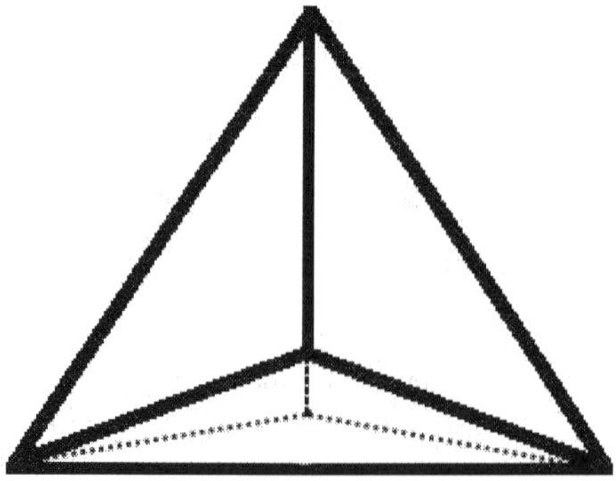

The center point of the base face makes a vertical line with the tip of the pyramid (the apex) where all solid line segments (slant edges) meet, making these line segments equal (base face line segments to the center point that are shown dashed in the figure). Since all angles in the base face are equal (because the base face sides are equal), each of them measures 60 degrees. The three dashed lines meeting at the center of the base face make 3 identical isosceles triangles, each such triangle with two 30-degree angles on the sides (not its central angle that measures 60 degrees).

We must find the value of one of these three dashed line segments drawn to the center of the base face, because it will act as a base for the right triangle we will create – with hypotenuse that will be the longest slant height of the pyramid (slant edge) and vertical height to the tip (our goal to find). To find the dashed line, we use the horizontal distance that the dashed line covers until reaching the center of the base face, and this is half of the isosceles triangle base ($\frac{11}{2}$), making the base 5.5 for the small right triangle we created. Now the dashed line we need is a small hypotenuse for this triangle. Since the angle between the small base and dashed line measures 30 degrees, we can easily find the hypotenuse (dashed line): $\cos(30^o) = \frac{5.5}{hyp}$, making the hypotenuse (dashed line) equal 6.350852961. Now we are ready to think of this dashed line as our base for the right triangle we build pointing up to the tip of the pyramid: the hypotenuse for this right triangle will be now the longest slant height (11), and vertical height of this right triangle is what we need.

We can use the *P. T.* to find the vertical height: $h_v = \sqrt{11^2 - 6.350852961^2} =$ 8.98146239. We also need to find the height of the base face (using 5.5 and 11), again we use the *P. T.* for this: $h_b = \sqrt{11^2 - 5.5^2} = 9.526279442$. Base face area is then $\frac{9.526279442(11)}{2} = 52.39453693$. Surface area is simply 4 times this amount (because all four faces are identical), giving 209.5781477 square feet. In square inches, this is about 30,179. The volume is then $\frac{(base\ face\ area)(vertical\ height)}{3} = \frac{52.39453693(8.98146239)}{3} = 156.8598543$ cubic feet. In cubic yards (1 cubic yard = 27 cubic feet), this is $\frac{156.8598543}{27} \approx 6$.

Answer: C

84. In the following figure, two mutually perpendicular images show an ellipsoid that perfectly encloses a closed right cylinder that shares the center point with the ellipsoid. The second image shows one shaded circle inscribed into an unshaded circle. The largest principal radius of the ellipsoid is 12 feet, the height of the cylinder is 15 feet, and the volume of the ellipsoid is 192π cubic feet. Find the surface area (in square inches) and volume (in cubic feet) of the cylinder.

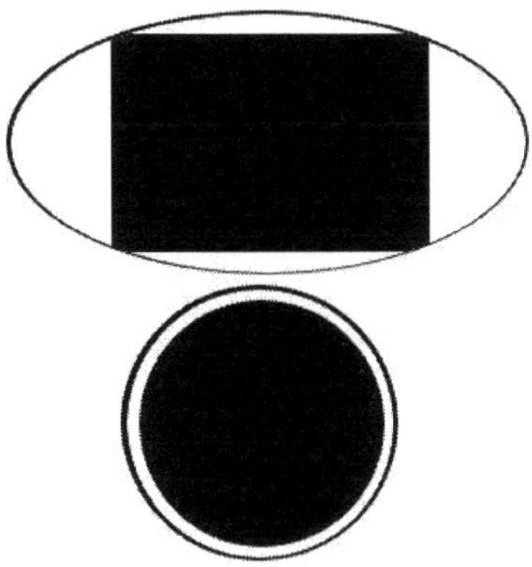

A. S_A = 33,625 in ^{2}, V = 8 ft ^{3}

B. S_A = 33,625 in ^{2}, V = 221 ft ^{3}

C. S_A = 234 in ^{2}, V = 381,704 ft ^{3}

D. S_A = 381,704 in ^{2}, V = 234 ft ^{3}

In this example we can imagine a football that matches the description (and image) of an ellipsoid. Since the second image (perpendicular to the first) is a circle, this means that the two of the three principal radiuses of the ellipsoid will be identical.

The largest principal radius of the ellipsoid is the horizontal radius in the first image, and the vertical smaller principal radius (in the first image) is one of the identical two radiuses in the second image. We need to find one of the identical radiuses r_i of the ellipsoid. Since the volume of the ellipsoid is 192π (and the volume formula is similar to sphere but the radiuses may vary as in the case here), this means that 192π = $(\frac{4\pi}{3})(12)(r_i^{\,2})$, meaning that r_i = 3.464101615 (each of the two smaller identical principal radiuses of the ellipsoid). We can now use the concept of direct proportion, because we know the large radius, small radius, and cylinder height half that is $\frac{15}{2}$ = 7.5. We can use these numbers to find the radius of the cylinder using direct proportion (see figure).

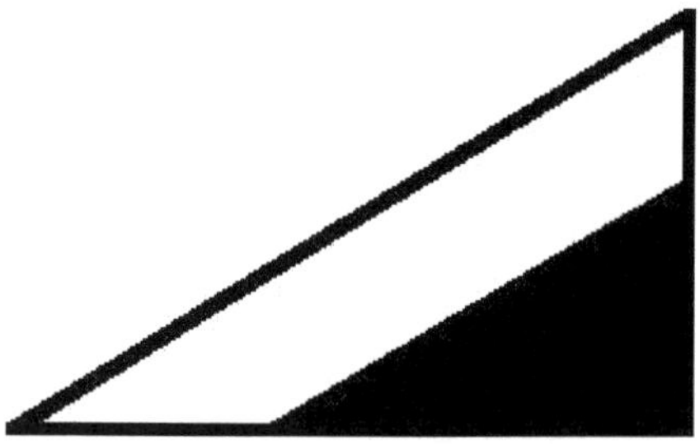

Cylinder height half along with the cylinder radius (black triangle) both show proportional decrease when compared to large horizontal principal radius of the ellipsoid along with the vertical principal (smaller) ellipsoid radius. Thus, their triangles are similar. Therefore, the proportion is: (ellipsoid small radius / ellipsoid large radius) = (cylinder radius r / cylinder height half), and this is $\frac{3.464101615}{12} = \frac{r}{7.5}$, and after cross multiplying we find the radius of the cylinder r = 2.165063509. Surface area of the closed cylinder is then 2π(2.165063509^2) + 2π(2.165063509)(15) = 233.5048595 square feet. In square inches this is about 33,625. Volume of the cylinder is $\pi(2.165063509^2)(15) = 220.8932334$ cubic feet.

Answer: B

85. In the following rectangular prism, the length (facing the reader) is 3 times the height; the width is 2 times the height. If the longest diagonal of the prism is 54 meters, what is true of the angles *x*, *y* and *z*?

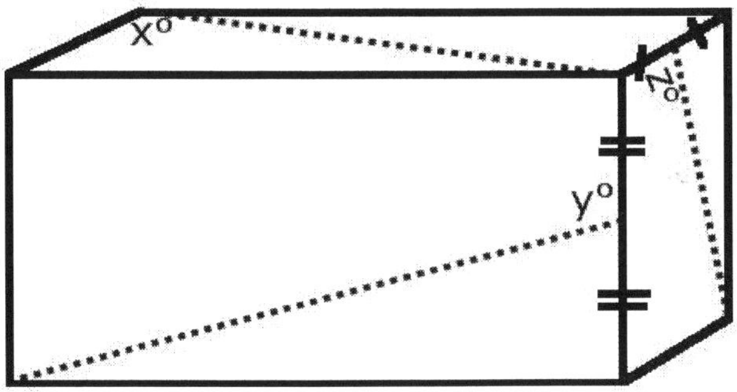

A. two of the angles are acute

B. the average of their measures is exactly 96 degrees

C. ($m\angle x + m\angle y$) > ($m\angle z + 20$)

D. ($m\angle z - 90$) = $9 + (\frac{9}{14})m\angle x$

Since $54 = \sqrt{l^2 + w^2 + h^2}$, and since $l = 3h$ and $w = 2h$, then $54 = \sqrt{9h^2 + 4h^2 + h^2}$ $= \sqrt{14h^2} = h\sqrt{14}$, and so height $h = 14.43210706$, $w = 28.86421413$, and $l =$ 43.29632119. Angle x measures $\tan^{-1}(\frac{43.29632119}{28.86421413}) = 56.30993247$ degrees. To find the measure of angle y, we find measure of the adjacent angle first: $\tan^{-1}(\frac{43.29632119}{\frac{14.43210706}{2}}) =$ 80.53767779 degrees, making measure of angle $y = 180 - 80.53767779 =$

99.46232221 degrees. Again, for angle z we find the adjacent angle first: $\tan^{-1}\left[\frac{14.43210706}{\frac{28.86421413}{2}}\right] = 45$ degrees, making angle $z = 180 - 45 = 135$ degrees.

We can eliminate the first choice, because only angle x is acute. The average of the three angles is $\frac{135 + 56.30993247 + 99.46232221}{3} = 96.92408489$ degrees, not exactly 96 degrees (it actually rounds to 97 degrees). The sum of angles x and y is 155.7722547, which is in fact greater than the sum of 20 and measure of angle z, which is 155.

Answer: C

86. The following figure shows a closed right cylinder that is perfectly inscribed in a sphere that shares the center point with the cylinder. The diameter of the cylinder is the same as its height, surface area of the sphere is 236π square meters, and the volume of the sphere outside the volume of the cylinder is 506π cubic meters. Find the measure of composite angle ($x + y$), and the surface area (in square feet) of the cylinder.

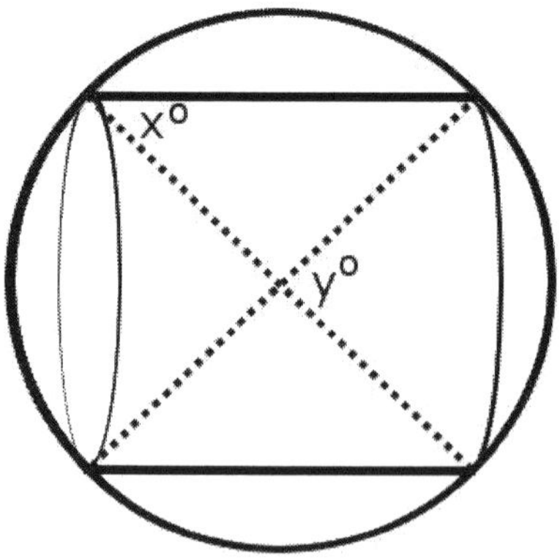

A. 112.5°, $S_A = 253$ ft ^{2}

B. 180°, $S_A = 2{,}721$ ft ^{2}

C. 135°, $S_A = 2{,}721$ ft ^{2}

D. 155°, $S_A = 253$ ft ^{2}

Since the enclosed cylinder has diameter that is equal to its height, it must be a cube-shaped cylinder. Now, the cylinder center point is shared with the sphere center point, therefore the dashed diagonals of the cylinder meet at this center point, making each half of the cylinder diagonal (there are four such diagonal halves) equal the radius of the sphere.

Since the surface area sphere is 236π = $4\pi r_S^2$, the radius of the sphere r_S = 7.681145748. The volume of the sphere outside of the volume of the cylinder is given as 506π, which means that we can find the volume of the cylinder by subtracting the volume of the sphere outside the volume of the cylinder from the volume of the sphere. The volume of the sphere is $(\frac{4\pi}{3})(7.681145748^3) = 1{,}898.307776$, making the volume of the cylinder 1,898.307776 – 506π = 308.6618935. Since the cylinder height is the same as its diameter, the radius of the cylinder can be expressed as $2r_C = h$, and we will have the volume as $308.6618935 = \pi r_C^2(2r_C) = 2\pi r_C^3$, making the cylinder radius r_C = 3.662416365. Cylinder height $h = 2(3.662416365) = 7.32483273$. The cylinder surface area is then $2\pi(3.662416365^2) + 2\pi(3.662416365)(7.32483273) = 252.8346284$ square meters. In square feet this is 252.8346284(10.76391042) ≈ 2,721.

Since the cylinder diagonals make four identical isosceles triangles (because the 2-dimensional view of the cylinder is a square with diagonals making four 90-degree angles with each other at the center point), and each isosceles triangle has a 90-degree angle, this makes the two other angles in the isosceles triangle measure $\frac{180-90}{2} = 45$ degrees each. Thus, angle y measures 90 degrees, angle x measures 45 degrees, and their measure sum is 135 degrees.

Answer: C

87. The following figure shows a trapezoidal prism. Find the measure of angle between the two bold (dashed) diagonals.

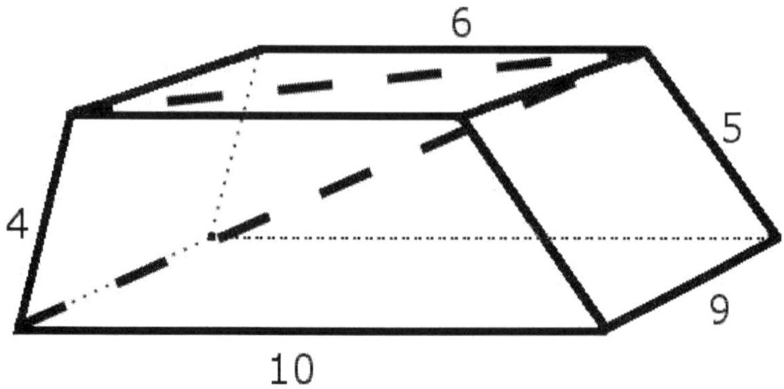

A. 97°

B. 64°

C. 161°

D. 19°

We see clearly that the diagonals along with a slanted height of 4 make a triangle. With this in mind, we need to find the lengths of the two diagonals and then use the *Law of Cosines* to find the angle between them. We will name the short diagonal *A* and long diagonal *B*. Since the components of the *A* diagonal are 6 and 9 (why?), we have that the *A* diagonal length is $\sqrt{6^2 + 9^2}$ = 10.81665383. The *B* diagonal is more difficult to find, because we need to figure out the portion of base 10 that it covers (this is one of the three components, the other two are 9 and height *h*, because this is a 3-dimensional diagonal). First, we find the vertical height *h*. To do that, we can use the given slanted heights and solve for *h* in the following way. Notice that there are two imaginary triangles with slanted heights 4 and 5 (hypotenuses) that make the figure a trapezoidal prism apart from a regular rectangular prism with dimensions *h*, 6 and 9. In other words, the chunks of the base 10 are 6 and two unknown values surrounding the 6 value. We will label the left chunk *x* and the right chunk *y*, so that $10 = x + y + 6$, which makes $y = 4 - x$.

Now, using the left imaginary triangle with hypotenuse 4, we have $4^2 = h^2 + x^2$, and using the right imaginary triangle with hypotenuse 5 we have $5^2 = h^2 + y^2$. Now, using the equation $y = 4 - x$, this becomes $5^2 = h^2 + (4 - x)^2 = h^2 + 16 - 8x + x^2$. Now, we can isolate the h^2 on one side for the two equations we had for the two imaginary triangles and then set those expressions equal as follows: equation with hypotenuse 4 becomes

$h^2 = 4^2 - x^2$, and equation with hypotenuse 5 becomes $h^2 = 5^2 - x^2 + 8x - 16$. Setting the two equations equal (because they both represent h^2) we have $4^2 - x^2 = 5^2 - x^2 + 8x - 16$. Upon simplifying and solving for x, we get $x = \frac{7}{8}$. This makes height $h = \sqrt{16 - \left(\frac{7}{8}\right)^2} = 3.903123749$. We have found the second component for the B diagonal. Also, the portion of the base 10 we need (the third component of the B diagonal) is $x + 6 = \frac{7}{8} + 6 = 6.875$.

We now have all three components to find the B diagonal, which is $\sqrt{6.875^2 + 9^2 + 3.903123749^2} = 11.97914855$. We now have all the information for the triangle with sides 4 (slant height in the figure), 11.97914855 (B diagonal length) and 10.81665383 (A diagonal length) in order to use the *Law of Cosines* to find the measure of angle C between the A and B diagonals. We have $4^2 = 10.81665383^2 + 11.97914855^2 - 2(10.81665383)(11.97914855)(\cos(C))$, so that $C = \cos^{-1}(0.943474108) \approx 19$ degrees.

Answer: D

88. The following figure shows a rectangular oblique pyramid whose base is a square. All other faces are distinct in length (there is no right angle anywhere except between the sides of the base). Pyramid vertical height *EF* is also shown.

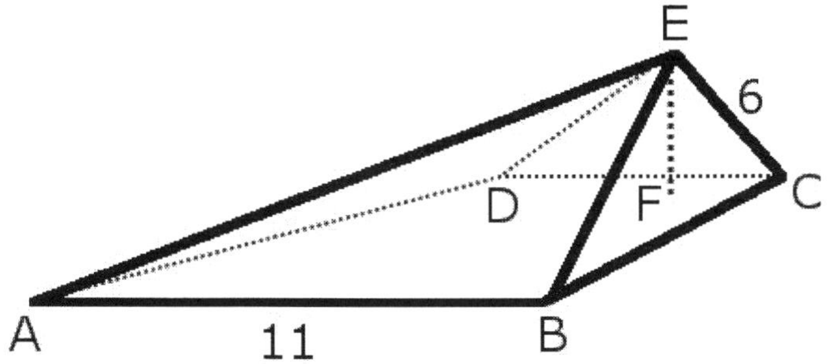

Find the volume and surface area of the pyramid using the following information:

m<*DCE* = 52° , m<*EBC* = 32°, m<*BCE* = 79°, m<*ABE* = 81°, m<*ABF* = 79°, area of triangle *ABF* is 49.

A. V = 284, S_A = 269

B. V = 269, S_A = 284

C. V = 259, S_A = 294

D. V = 294, S_A = 259

To find the surface area of the pyramid, we need to find the areas of all the faces that make up the pyramid, and then find their sum. The base face is 11(11) = 121. We will now work with triangle face EBC. To find its area, we first find its height. Since angle BCE measures 79 degrees, we have $\sin(79^o) = \frac{h}{6}$, so that h = 5.889763101. Area of triangle face EBC is then $\frac{11(5.889763101)}{2}$ = 32.39369705.

Now, we will work on triangle face ABE. To find its height, we first need to find the length of side BE. Using *The Law of Sines* on triangle face EBC, we have $\frac{\sin(79^o)}{BE} = \frac{\sin(32^o)}{6}$, so that BE = 11.11445365. Since $_m$<ABE = 81 degrees, we have $\sin(81^o) = \frac{h}{11.11445365}$, so that height h for triangle ABE is 10.97761628. Area of triangle face ABE is then $\frac{11(10.97761628)}{2}$ = 60.37688955. Now, working with triangle face CDE, we see that the height h for it is: $\sin(52^o) = \frac{h}{6}$, so that h = 4.728064522. Then the area of triangle face CDE is $\frac{4.728064522(11)}{2}$ = 26.00435487.

Now working with the last triangle face ADE, we see that we need to find its height h and length AE in order to find its area. Since angle BCE measures 79 degrees, angle ADE also measures 79 degrees (why?). We now have a triangle ADE with AD = 11 and measure of angle ADE is 79 degrees. In order to find base AE for triangle ADE, we need to find side ED first. To find ED, we use the triangle CDE with CD = 11, $_m$<DCE = 52°, and EC = 6. It is clear that we can use the *Law of Cosines* to find DE, so that $DE^2 = 11^2 + 6^2 - 2(11)(6)(\cos(52^o))$ = 75.73268526, so that DE = $\sqrt{75.73268526}$ = 8.70245283.

Now, since we know DE we can find base AE for triangle ADE using the *Law of Cosines* again, so that $AE^2 = 11^2 + 8.70245283^2 - 2(11)(8.70245283)(\cos(79^\circ)) = 160.2015471$, so that $AE = \sqrt{160.2015471} = 12.65707498$. Now, we are missing height *h* for triangle face ADE. This height will start from E and will be perpendicular to AD. We can create an imaginary right triangle with hypotenuse ED = 8.70245283 and angle whose measure is 79 degrees (the opposite side of which is the height *h* we need to calculate). We have $\sin(79^\circ) = \frac{h}{8.70245283}$, so that *h* = 8.542564261. Area of the triangle ADE is then $\frac{12.65707498(8.542564261)}{2} = 54.06193819$. Finally, the surface area of the pyramid is the sum of the areas of its five faces we found, that is, 54.06193819 + 26.00435487 + 60.37688955 + 32.39369705 + 121 ≈ 294.

To find the volume of this pyramid, we need to find its height EF (a line from the apex, or tip of the pyramid, that is perpendicular to the base face – indicated as a dashed red line in the figure). Since the area of the triangle ABF is 49, we have $\frac{11h}{2} = 49$, so that *h* = 8.909090909. Note that this triangle lies on the base face plane. We can create an imaginary right triangle on the same plane that has the dimensions: height we just found, hypotenuse BF and angle ABF that measures 79 degrees. BF is then: $\sin(79^\circ) = \frac{8.909090909}{BF}$, so that BF = 9.075839646.

To find height EF of the pyramid, BF will now be part of another right triangle with hypotenuse BE = 11.11445365. We then have $EF = \sqrt{11.114453652^2 - 9.075839646^2} = 6.415622702$. Volume of the pyramid is $\frac{(base\ face\ area)(height)}{3} = \frac{(121)(6.415622702)}{3} \approx 259$.

Answer: C

89. In the following figure of the closed rhombic prism, four of the six faces are squares with side 5 feet. Determine the volume (in cubic inches) of the prism, and find the value of the longest diagonal.

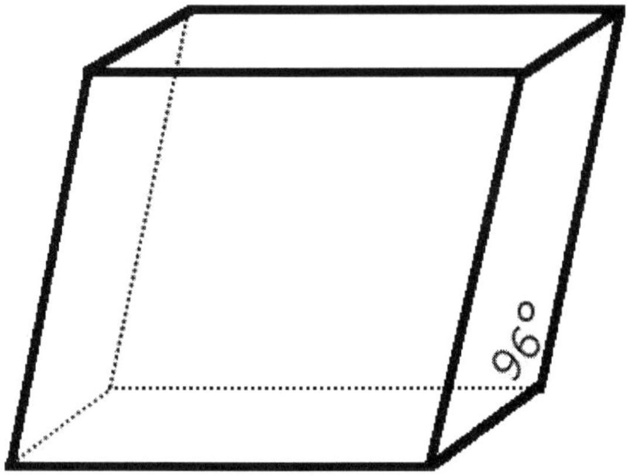

A. V = 214,817 in ^{3}, d = 9 ft

B. V = 214,817 in ^{3}, d = 7 ft

C. V = 124 in ^{3}, d = 9 ft

D. V = 124 in ^{3}, d = 7 ft

To find the volume, we first need to find the prism's height. The slant height of the prism will be the hypotenuse of an imaginary triangle that we need to use, and a 96-degree angle means that the adjacent angle measures 180 – 96 = 84 degrees. See the figure below.

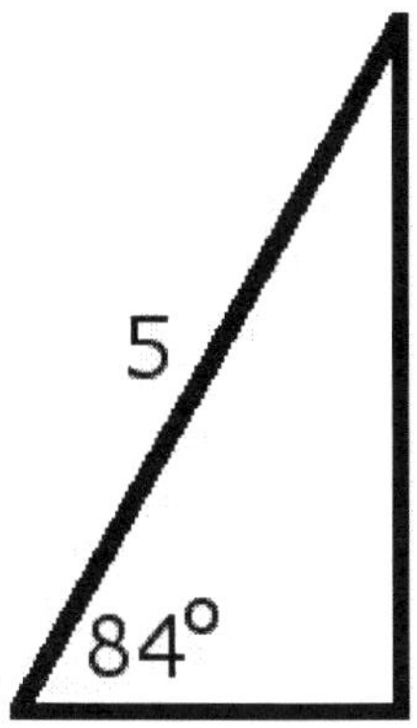

This angle is the second largest angle in the imaginary triangle (also this angle is adjacent to the hypotenuse). Hypotenuse of the triangle is 5, because the rhombic prism has all equal dimensions. Thus, our desired vertical height h is: $\sin(84^{\circ}) = \frac{h}{5}$, making h = 4.972609447. It follows that the volume of the prism is (5)(5)(4.972609447) = 124.3152362 cubic feet. Converting to cubic inches, we get (124.3152362)(1728) ≈ 214,817.

Now, to find the longest diagonal (there are two such diagonals, why?), we need to first imagine that we the prism slanted heights became all vertical (while keeping in mind that the slanted height is greater than the vertical height we found, so we must use 4.972609447 and **not** 5). The longest diagonal of the vertical-height prism would be $\sqrt{5^2 + 5^2 + 4.972609447^2}$ = 8.644469024. Now, we see that the rhombus made a horizontal shift to the right (it seems like it was stretched to the right), which means that the longest diagonal of the rhombic prism will definitely be longer than the one for the vertical-height prism. See the figure below.

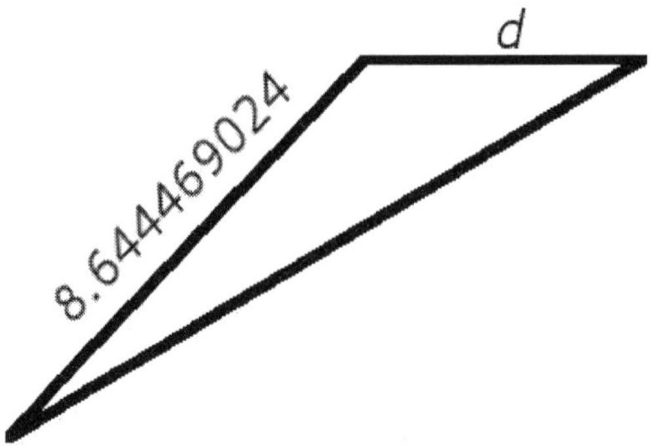

We need to find the horizontal distance *d* that the rhombus stretched to the right (this distance is also the small base of the right triangle in the first figure), so we use the same imaginary triangle once more for that: $\cos(84^\circ) = \frac{d}{5}$, giving the distance d = 0.522642316.

Now, we can use a triangle (shown in the last figure) consisting of sides 0.522642316 (horizontal stretch distance that we just found), 8.644469024 (vertical-height prism diagonal), and the desired rhombic prism diagonal. This now obviously implies that we can use the *Law of Cosines* to find the rhombic prism diagonal.

The only piece of information missing for it is the angle between the horizontal shift distance (0.522642316) and vertical-height prism diagonal (8.644469024). This angle is the one whose opposite side is the desired rhombic prism diagonal. We see that this angle must measure 135 degrees (because we had the original 45-degree angle between the vertical-height prism diagonal and its square side, and then added the available adjacent 90-degree angle because this square side got shifted to the right). Using the law to find the rhombic prism diagonal D, we get $D^2 = 8.644469024^2 + 0.522642316^2 - 2(8.644469024)(0.522642316)(\cos(135^\circ))$, and taking the square root of this result gives $D \approx 9$ feet.

Answer: A

90. The following figure shows a triangular prism with base 4 feet and length 9 feet (the triangle face has distinct dimensions). If the volume of the prism is 112,320 cubic inches and the shortest slant height is 3.8 feet, find the measure of the composite angle (*a* – *b*), and the longest diagonal of the prism (rounded to the nearest 0.1).

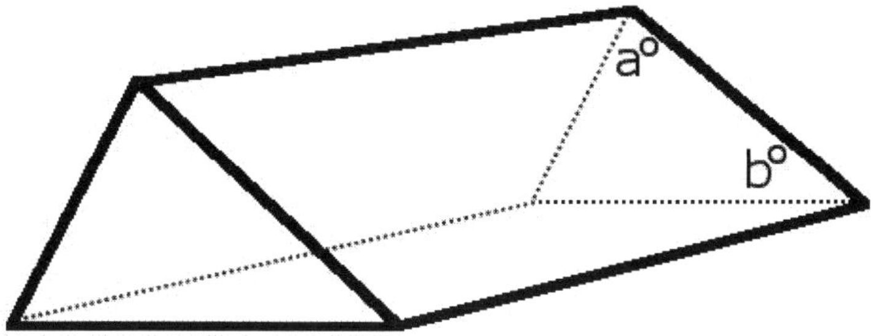

A. 4°, 10.1 ft

B. 22°, 10.1 ft

C. 4°, 9.8 ft

D. 22°, 9.8 ft

We assume that since the measure of composite angle $a - b > 0$, then $a > b$. The volume of the prism is $\frac{blh}{2} = 112{,}320$. The volume 112,320 is in cubic inches, so converting to cubic feet we have $\frac{112{,}320}{1728} = 65$ cubic feet. Then, from given information we have $65 = \frac{(9)(4)h}{2}$, and we get vertical height $h = 3.611111111$.

Now, the shortest slant height (adjacent left side of angle *a*) is 3.8, we can use this information to find a part of measure of angle *a* (thus splitting it). This part will be $\cos^{-1}\left(\frac{vertical\ height}{hypotenuse}\right) = \cos^{-1}\left(\frac{3.611111111}{3.8}\right) = 18.14114266$ degrees. To find the remaining part of measure of angle *a*, we must first find the horizontal distance *d* that the shortest slanted height covers (a small part of the prism base), so we use the part of angle *a* we just found: $\tan(18.14114266^\circ) = \frac{d}{3.611111111}$, which gives $d =$ 1.183163786.

We now know the part of the base (horizontal distance) that the longer slant height covers (the adjacent side for both angles *a* and *b*): $4 - 1.183163786 = 2.816836214$.

We can then find the remaining part of angle *a*, angle *b* and the longest slanted height itself. The remaining part of measure of angle *a* is: $\tan^{-1}(\frac{2.816836214}{3.611111111}) = 37.95590339$ degrees. Then angle *a* measures 37.95590339 + 18.14114266 = 56.09704605°. Angle *b* should measure less than this amount, so we find it using adjacent and opposite sides of angle *b*: $\tan^{-1}(\frac{3.611111111}{2.816836214}) = 52.04409661$ degrees, which agrees with the fact that $a > b$. Thus, $(a - b) = 56.09704605° - 52.04409661° \approx 4$ degrees. The longer slant height *s* can be found by using angle *b*: $\cos(52.04409661°) = \frac{2.816836214}{s}$, giving $s =$ 4.579813277.

The longest diagonal of the prism will be determined by using the largest two dimensions of the prism. In this case, this is the longest slanted height we just found and the given length. Thus, the longest diagonal of the prism is $\sqrt{4.579813277^2 + 9^2}$ $\approx$ 10.1 feet. As a side note, we could have found the angle measures of *a* and *b* by first finding the third supplementary angle to them. You may do so for your own practice.

Answer: A

91. In the following figure, the area is 50. If segments *BD* = 5, *BE* = 7, find the small dashed height and segments *AD* and *DC*.

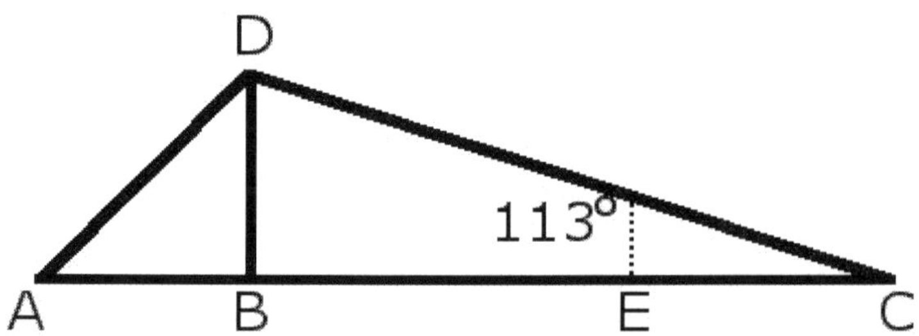

A. $h = 2$, $AD = 8$, $DC = 12$

B. $h = 2$, $AD = 13$, $DC = 10$

C. $h = 2$, $AD = 10$, $DC = 13$

D. $h = 2$, $AD = 10$, $DC = 8$

We can subtract 90 degrees from 113 degrees to find the measure of the acute angle that remains: this angle will also be equal to angle C (why?), so that measure of angle C = 113 – 90 = 23 degrees. Knowing that BD = 5, we can find EC: we have $\tan(23^o) = \frac{5}{7 + \text{EC}}$, making 7 + EC = 11.77926183, so that EC = 4.779261829.

Notice that triangle BDC and the small triangle (defined by the dashed vertical height and angle C) are similar, thus we can use direct proportion to find the dashed height h (using BC, EC and BD): $\frac{11.77926183}{5} = \frac{4.779261829}{h}$, and simplifying for h we get h = 2.028676286. Now, since the area is 50, we take into account the missing part of the large base of the figure, so we name it y. Then, $50 = \frac{(11.77926183 + y)(5)}{2}$, which gives y = 8.22073817. AD then can be found by *P. T.* as $\sqrt{8.22073817^2 + 5^2}$ = 9.62187799. DC is found using angle C and BD: $\sin(23^o) = \frac{5}{DC}$, making DC = 12.79652333.

Answer: C

92. In the following figure, a right triangle is inscribed in a circle. The area of the circle is 2.125 as large as its circumference. The area of the triangle is $\frac{11}{40}$ the area of the circle. Hypotenuse passes through the center of the circle. Find the measure of the second smallest angle of the triangle, arc length encompassed by this angle (to the nearest 0.01), and area of the circle region labeled *x* (to the nearest 0.01).

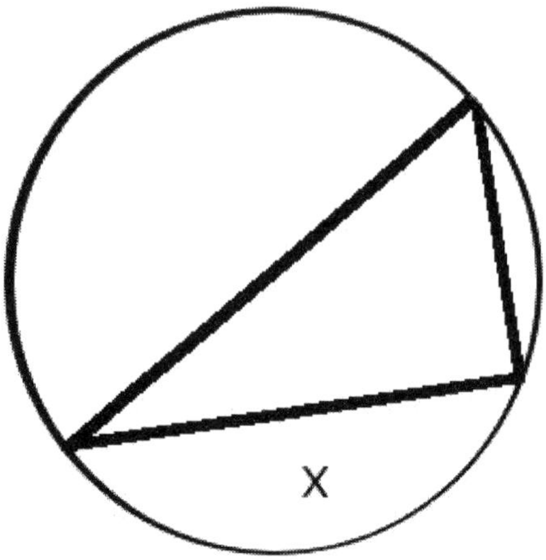

A. 30°, $L_A = 4.43$, $A = 8.92$

B. 30°, $L_A = 4.43$, $A = 11.15$

C. 60°, $L_A = 11.15$, $A = 8.92$

D. 60°, $L_A = 8.92$, $A = 11.15$

Since the area of the circle is 2.125 the circumference of the circle, $\pi r^2 = 2.125(2\pi r)$, which means that the radius r is 4.25.

The area of the circle is then $\pi(4.25^2) = 56.74501731$. The area of the enclosed right triangle is $\frac{11}{40}$ of this amount, that is, 15.60487976. Since the hypotenuse passes through the center of the circle, it must be that the hypotenuse is equal to the circle diameter, that is, $2(4.25) = 8.5$. We need to first find the second smallest angle in this triangle. See the figure below.

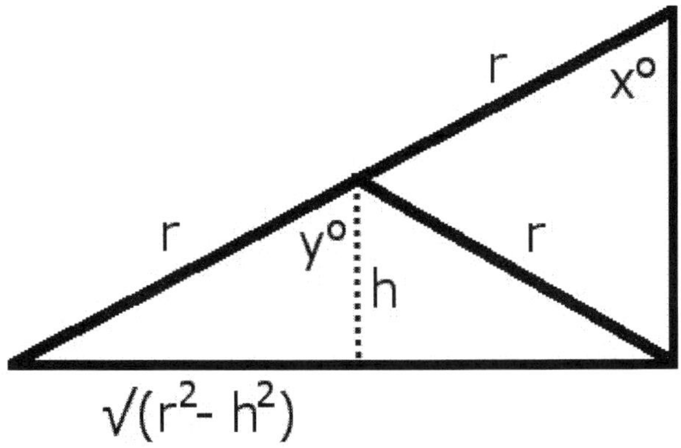

We need to find measure of angle *x*. We could simplify the problem by working with angle *y*, which is congruent to angle *x* (why?). So we built a small right triangle with angle *y* (which starts right at the center of the circle) as a similar triangle to the parent large triangle, so its hypotenuse will be exactly half the original hypotenuse - it will be equal to the circle radius of 4.25. Notice that two isosceles triangles are formed that make up the large parent triangle. We will label the dashed height of the small right triangle as *h*. The base of this right triangle must be $\sqrt{4.25^2 - h^2}$. Thus, the base of the parent triangle (original triangle in the figure) is twice this amount (why?), namely $2\sqrt{4.25^2 - h^2}$, and height of the parent triangle is 2*h* (why?). Since we know the area of the parent triangle, we have 15.60487976 = $\frac{2h\left(2\sqrt{4.25^2 - h^2}\right)}{2}$, which upon simplifying and getting rid of the square root gives a quadratic equation $h^4 - 4.25^2 +$ 60.87806808 = 0. Using quadratic formula, we get $h^2 = \frac{(4.25)^2 - 9.096242847}{2} =$ 4.483128577, so that h = 2.117339977. We do not use the second root, because the height *h* should be less in value than the base. Angle *y* (and also angle *x*) measures $\cos^{-1}(\frac{2.117339977}{4.25})$ = 60.1191716 degrees.

To find the length of the arc *A* encompassed by this angle, we can use the fact that the ratio of this angle to 180 degrees will be equal to the ratio of the encompassed arc to the entire circle circumference. Thus, we have $(\frac{60.1191716^{\circ}}{180^{\circ}}) = \frac{A}{8.5\pi}$ which gives arc length *A* = 8.918858648. To find the area of the circle region labeled in the original figure, we subtract the area of the large isosceles triangle (which has the largest angle

2*y*, height *h* and base that is also the base of the original parent triangle) from the area that includes this isosceles triangle and the region in question.

The area that includes the isosceles triangle and region in question is found by using the fact that the ratio of angle 2*y* to 360 degrees will equal to the ratio of the area that includes the area of isosceles triangle and region marked *x* in the original figure to the entire area of the circle. Measure of angle 2*y* is 2(60.1191716°) = 120.2383432°. The ratio is then $\frac{120.2383432^o}{360^o}$ = 0.333995398. Thus, the area of the circle region that includes region *x* in the original figure and the isosceles triangle is 0.333995398(56.74501731) = 18.95257464. Since the base of the isosceles triangle is $2\sqrt{4.25^2 - h^2}$, which is 7.370039734, the area of this triangle is 7.370039734 • $\frac{2.117339977}{2}$ = 7.80243988. Thus, the desired region marked *x* in the original figure is 18.95257464 – 7.80243988 ≈ 11.15.

Answer: D

93. The following 7-sided polygon (with identical sides) is perfectly inscribed in a circle. If the circumference of the circle is 18π feet, find the length (in yards) of the dashed line.

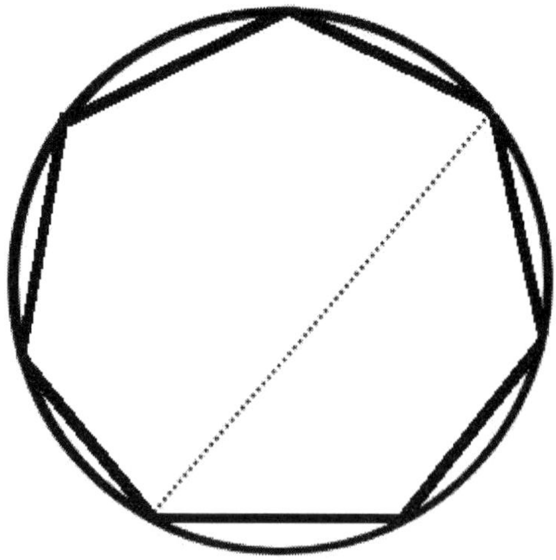

A. 18 yd

B. 3 yd

C. 9 yd

D. 6 yd

Since the circumference of the circle is 18π feet, the radius *r* is 9 feet. Now, there are seven identical triangles that can be drawn from the center of the circle to each of the seven edges of the polygon. The base of each of those triangles will be the side length of the polygon. Since there are seven identical triangles that make up this polygon, each of the central angles (largest angle for each of the seven identical triangles) will measure $\frac{360}{7}$ = 51.42857143 degrees. Notice that we can connect the dashed line with the two circle radiuses that lie directly on top of it, thus making an isosceles triangle. This is done by drawing the two radiuses above so that they terminate in the same two edges of the polygon that the dashed line connects. See the figure below showing the isosceles triangle we now created.

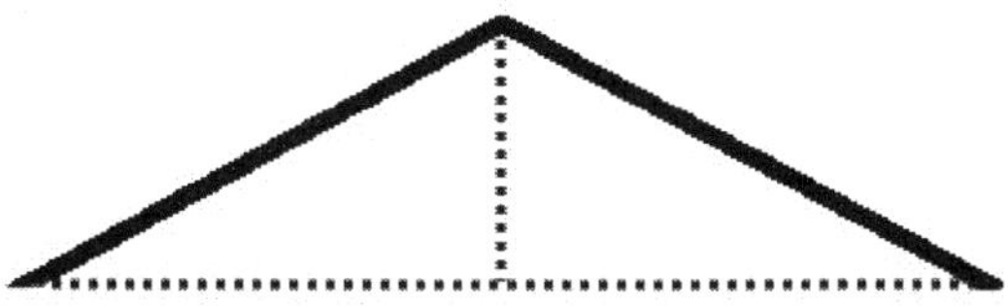

The base *B* of this triangle is the dashed line we need to find. The identical two sides of the isosceles triangle are both equal to the radius of the circle, which is 9. Measure of the largest angle of this triangle will be the sum of measures of three identical central angles of the identical seven triangles that made up the polygon, that is, 3 • 51.42857143° = 154.2857143 degrees. We can divide the isosceles triangle in half, creating two identical right triangles, with the second largest angle in each triangle thus measuring $\frac{154.2857143}{2}$ = 77.14285715 degrees. Thus, the base *b* for any of the two small right triangles created is sin(77.14285715°) = $\frac{b}{9}$, making *b* = 8.77435121, and subsequently *B* = 2*b* = 17.54870242 feet. This is the distance of the dashed line (large base for the isosceles triangle) we were asked to find. Converting the answer to yards (1 yard = 3 feet), we get the distance to be about 6 yards.

Answer: D

94. The following figure shows a triangular prism. The height is 3 times shorter than the length, and the width is 3 times shorter than the height. Diagonals are also labeled.

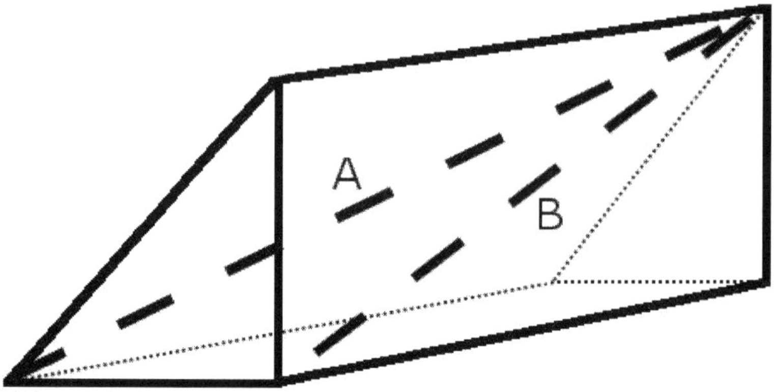

The following statements are delivered:

I. The sum of the diagonals *A* and *B* is less than twice the base face diagonal (not shown)

II. The product of the base face and *B* diagonals is less than the square of the *A* diagonal

III. The product of the slant height and the base face diagonal is less than the sum of the diagonals *A* and *B*

IV. Measure of the angle between the width and slant height is greater than 12 times the measure of the acute angle between the diagonals *A* and *B*

Which statements are FALSE?

A. I, III, and IV

B. I and II

C. I, II and IV

D. II and IV

If we let the length to be x, then the height is $\frac{x}{3}$. Width is then $\frac{\frac{x}{3}}{3}$, or $\frac{x}{9}$.

Diagonal A is determined by the height, length and width, so it is $\sqrt{x^2 + \left(\frac{x}{3}\right)^2 + \left(\frac{x}{9}\right)^2} =$ $1.059932446x$. The B diagonal is determined by length and height, so it is $\sqrt{x^2 + \left(\frac{x}{3}\right)^2}$ $= 1.054092553x$. The base face diagonal (which is not drawn in the figure) is calculated using width and length: $\sqrt{x^2 + \left(\frac{x}{9}\right)^2} = 1.006153904x$. The sum of diagonals A and B is then $1.059932446x + 1.054092553x = 2.114024999x$. Twice the base face diagonal is $2(1.006153904x) = 2.012307808$, which is not greater than the sum of A and B diagonals. This makes statement I false.

The product of base face and B diagonals is $1.006153904x \bullet 1.054092553x = 1.060579337x^2$. The square of the A diagonal is $(1.059932446x)^2 = 1.12345679x^2$. This amount is greater than the product of base face and B diagonals, thus making statement II true. The slant height (hypotenuse for width and height) of the figure is $\sqrt{\left(\frac{x}{9}\right)^2 + \left(\frac{x}{3}\right)^2} = 0.351364184x$. The product of this and the base face diagonal is $0.351364184x \bullet 1.006153904x = 0.353526446x^2$. The sum of A and B diagonals is $2.114024999x$, as calculated earlier. Since $x > 0$, we see that even though for small values of x the statement III is true, however, for larger x values the statement III is false. Angle between the width and slanted height measures $\tan^{-1}\frac{\frac{x}{3}}{\frac{x}{9}} = \tan^{-1}(3) =$ 71.56505118 degrees. The angle between the A and B diagonals measures $\cos^{-1}\left(\frac{1.054092553x}{1.059932446x}\right) = 6.017285055$ degrees. Since $12(6.017285055) = 72.20742066^\circ >$ 71.56505118°, this makes statement IV false.

Answer: A

95. An unusually large football designed for Cyclops is a perfect example of an ellipsoid that has a large principal radius equal to 9 feet. Its two other principal radiuses are equal to each other. If the surface area of the football is 200 square feet, find the sum of all the hypotenuses (in meters) joining the principal radiuses in this ellipsoid.

A. 15 m

B. 50 m

C. 87 m

D. 26 m

Since $r_1 = 9$, $r_2 = r_3$, we can work with either one of the two unknown principal radiuses, or simply label r as one of the two unknown principal radiuses. The surface area is then

$$200 = 4\pi\left(\frac{(r^2)^{1.6}+(9r)^{1.6}+(9r)^{1.6}}{3}\right)^{\frac{1}{1.6}} =$$

$$4\pi\left(\frac{r^{3.2}+33.63473537r^{1.6}+33.63473537r^{1.6}}{3}\right)^{\frac{1}{1.6}} =$$

$4\pi\left(\frac{r^{3.2}+67.26947074r^{1.6}}{3}\right)^{\frac{1}{1.6}}$, which means that $(\frac{200}{4\pi})^{1.6} = \frac{r^{3.2}+67.26947074r^{1.6}}{3}$, so that $251.2079997 = r^{3.2} + 67.26947074r^{1.6}$, and rewriting the equation we have $r^{3.2} + 67.26947074r^{1.6} - 251.2079997 = 0$. Using quadratic formula, we have $r^{1.6} = \frac{-67.26947074 + 74.36406183}{2} = 3.547295545$, which means that $r = (3.547295545)^{\frac{1}{1.6}} = 2.206408177$. This is what the two other non-given principal radiuses measure in feet.

We can simplify the further work in finding the hypotenuses, noting that each plane containing two diameters (there are three such planes, why?) will have four identical hypotenuses joining its principal radiuses. There will be two identical planes and one remaining distinct plane (because two of the three principal radiuses are equal). Taking one of the two identical planes, one of the four identical hypotenuses in any of these two planes will be $\sqrt{9^2 + (2.206408177)^2} = 9.26651159$. In the third plane (where the two principal radiuses are equal), one of the four identical hypotenuses will be $\sqrt{(2.206408177)^2 + (2.206408177)^2} = 3.120332368$.

Thus, the sum of all the hypotenuses in the three planes (two planes with principal radiuses 9 and 2.206408177 and one plane with two identical principal radiuses that are 2.206408177, and noting that each plane has four identical hypotenuses) is 2 • 4 •

$9.26651159 + 4 \cdot 3.120332368 = 86.61342219$ feet. Converting to meters this sum is $86.61342219(0.3048) \approx 26$ meters.

Answer: D

96. The following figure shows a star inside a hexagon with six identical sides, each measuring 4 meters. What is the area of the star figure (in square kilometers)?

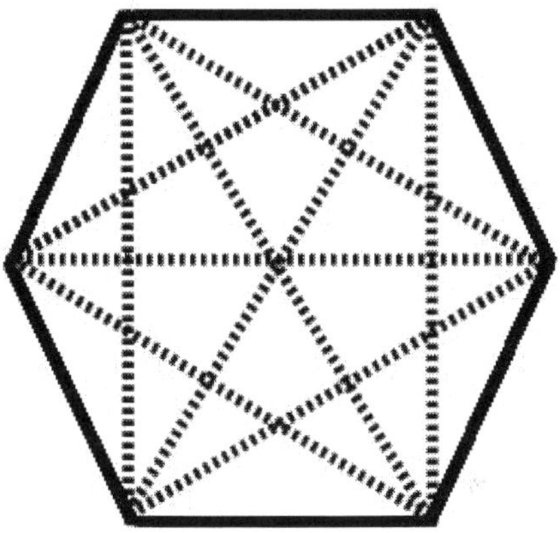

A. 28 km^2

B. $(2.77 \cdot 10^{-5})$ km^2

C. 14 km^2

D. $(1.39 \cdot 10^{-5})$ km^2

We can simplify the exhilarating work in calculating the area of this complex star figure by simply subtracting the hexagon regions not included in the star figure from the area of the hexagon. The region to be subtracted consists of six identical isosceles triangles, each with base equal to the hexagon side of 4. The total angle measure sum of the hexagon is $180(6 - 2) = 720$ degrees, so that each angle measures $\frac{720}{6} = 120$ degrees.

Now, it is easy to see that each of the six isosceles triangles has two equal angles which make them isosceles, and this angle must measure (120 – 90) = 30 degrees. One may see this when looking at the adjacent rectangle whose angles take 90 degrees from a 120-degree hexagon angle measure. To find the area of one of the six isosceles triangles, we first need to find its height. We split the isosceles triangle in half (creating two identical right triangles), so that its base becomes $\frac{4}{2}$ = 2. Since we know the measure of one of the angles already for one of these right triangles (it is 30 degrees), it must be that its height *h*: $\tan(30^\circ) = \frac{h}{2}$, so that *h* = 1.154700538.

Thus, the area of the isosceles triangle is $4 \bullet \frac{1.154700538}{2}$ = 2.309401077. Since there are six such isosceles triangles, the total area (their sum) is 6 • 2.309401077 = 13.85640646. Now, the area of the hexagon is $\frac{4^2(3\sqrt{3})}{2}$ = 41.56921938. Thus, the area of the star figure is 41.56921938 – 13.85640646 = 27.71281292 square meters. In square kilometers this is $2.77 \bullet 10^{-5}$.

Answer: B

97. The following figure provides a 3-dimensional coordinate illustration of a closed rhombic prism inscribed in a closed rectangular prism. If each side of the rhombic prism is 5 feet, find the internal surface area and volume of the two regions of the rectangular prism that do not include the rhombic prism (in square and cubic meters).

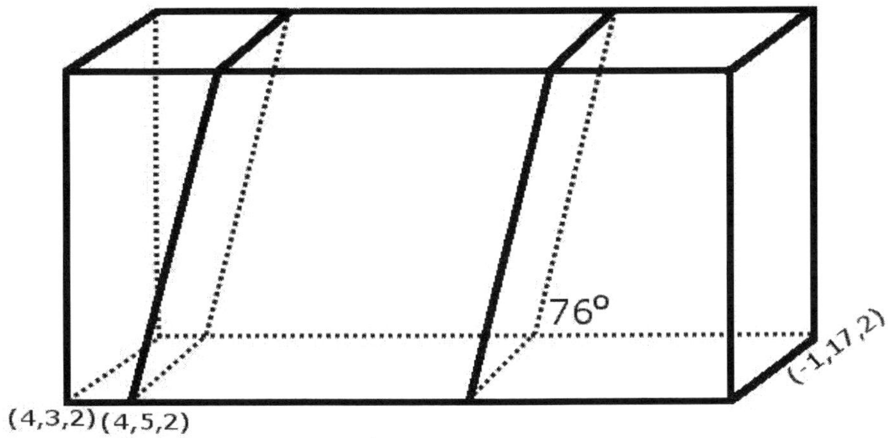

A. $S_A = 276\ m^2$, $V = 218\ m^3$

B. $S_A = 6\ m^2$, $V = 26\ m^3$

C. $S_A = 218\ m^2$, $V = 276\ m^3$

D. $S_A = 26\ m^2$, $V = 6\ m^3$

We will begin by working with the first region of the rectangular prism that does not include the rhombic prism (this is the left region adjacent to the rhombic prism) by first finding the vertical height of the rectangular prism.

The bottom side (label it $base_1$) of the first region has value of 2, since the coordinates of (4,3,2) and (4,5,2) make this obvious. The top side (base2) has value of base1 added with an additional chunk that we need to find. This chunk is found by using the slanted height (given as 5 that is a rhombic prism dimension) and angle of 76 degrees (angle between the slanted height and $base_2$ – why is it 76 degrees?). The chunk is then $\cos(76^{\circ}) = \frac{chunk}{5}$, so that the chunk is 1.209609478, and $base_2$ is then 1.209609478 + 2 = 3.209609478. The vertical height h of the prism is $\sin(76^{\circ}) = \frac{h}{5}$, which gives h = 4.851478631. The surface area of the left region consists of the face including $base_1$, $base_2$, vertical height, and slant height (two such faces), rhombic prism face (a square with side 5), a face that has dimensions $base_2$ and 5 (top face), face that consists of 5 and $base_1$ (bottom face), and a face that has dimensions 5 and vertical height. Thus the internal surface area of the left region is

$2\left\{(2)(4.851478631) + \frac{1.209609478(4.851478631)}{2}\right\}$ + 5(5) + 5(3.209609478) + 2(5) + 5(4.851478631) = 100.5797496.

For the second region outside the rhombic prism, we will again have two distinct bases for the two identical faces that have a slanted height included, and we will name these two bases $base_3$ (bottom base) and $base_4$ (top base). Since the total base length of the rectangular prism is 17 – 3 = 14 (we see that from the coordinates (-1,17,2) and (4,3,2)), $base_3$ will be 14 – 2 – 5 = 7. $Base_4$ is then 7 minus the amount which is equal to the chunk (1.209609478) we found earlier for the left region (why are these two chunks the same?), so $base_4$ is 7 – 1.209609478 = 5.790390522. The internal surface area for the second region is then $2\left\{\frac{1.209609478(4.851478631)}{2} + 5.790390522(4.851478631)\right\}$ + 5.790390522(5) + 5(5) + 7(5) + 5(4.851478631) = 175.2616521.

Hence, the total internal surface area of the first and second regions outside the rhombic prism is 175.2616521 + 100.5797496 = 275.8414017 square feet. In square meters this is 275.8414017(0.09290304) ≈ 26 square meters. The volume of the two regions combined is 2(4.851478631)(5) + $2\left\{\frac{1.209609478(4.851478631)(5)}{2}\right\}$ + 5.790390522(4.851478631)(5) = 218.3165384 cubic feet. In cubic meters this is about 6.

Answer: D

98. Find the measure of the smallest angle (in radians) in the triangle formed by the boundaries of the *y*-axis and the following equations:

$$y = 2x + 4.5$$

$$y = -3x - 3.5$$

A. 27 radians

B. 18 radians

C. 0.3 radians

D. 0.5 radians

The graph in question is shown below. The *A* line is $y = 2x + 4.5$, and the *B* line is $y = -3x - 3.5$.

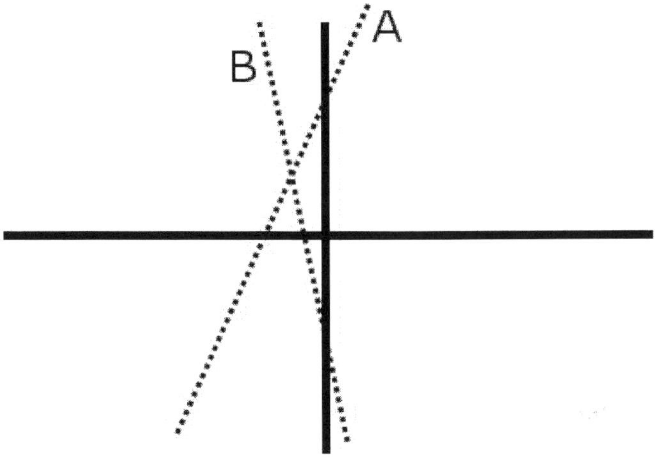

The way to determine the point of intersection of the two given equations is to first set their y-values to be equal, that is $2x + 4.5 = -3x - 3.5$, and solving for x we get $x = -\frac{8}{5}$. This is the x-value of the intersection of the two given lines. This also means that the y-value of their intersection (when plugging in the found x-value in either of the two equations) is $y = 1.3$. So, we can see that the third boundary for the triangle formed will be the y-axis (where $x = 0$).

Now, we can clearly see that the smallest angle candidate will not be the angle where the given equations meet, but rather the two angles that are formed with the y-axis (where $x = 0$). Since the *B* line (equation $y = -3x - 3.5$) is steeper, another words, has a steeper slope (why?), this means that angle formed with the y-axis by the *B* line will be smaller than the angle formed with the y-axis by the *A* line. One could assume that the *Law of Cosines* will be ideal for this problem, but it is simpler than that. We can get away with standard trigonometry in finding this angle measure. We can use the arctangent ($\tan^{-1}$) to find this angle, all we need is create an imaginary triangle with opposite and adjacent sides for the angle. The opposite side is the horizontal distance covered from the point $x = 0$ to the point where the two equations meet, that is, $\frac{8}{5}$ (in absolute value), and the adjacent side is the vertical distance between the point of intersection of the *B* line with the y-axis (where $y = -3.5$) and the point $y = 1.3$. This distance is therefore $3.5 + 1.3 = 4.8$ (in absolute value). Thus, the angle is question measures $\tan^{-1}(\frac{\frac{8}{5}}{4.8})$ = 18.43494882 degrees. Converting this answer to radians (1 degree = $\frac{\pi}{180}$ radians), we get 18.43494882($\frac{\pi}{180}$) ≈ 0.3 radians.

Answer: C

99. A car travels 45 miles per hour in a perfectly circular road (where starting in one point will eventually get the motorist to the same exact point). If the area bounded by this circular road is 670 square miles, find the rate of change of the measure of the central angle (in radians per minute, to the nearest 0.01) of this circular road as the car travels at this constant speed.

A. 3.08 rad/min

B. 2.94 rad/min

C. 0.03 rad/min

D. 0.05 rad/min

First we find the radius of the circular road, so that πr^2 = 670, so that radius r = 14.60368528 miles. The circumference of the circular road is 2π(14.60368528) = 91.75766078 miles. Since 1 hour = 60 minutes, we have 45 mi / 60 min = 0.75 mi/min as the speed of the vehicle. Now, we know that 2π radians takes 360 degrees, and we also know that 360 degrees will take the entire circumference of the circular road, so that the car covers 91.75766078 miles per 360 degrees. Thus, we have **{**(0.75 mi/min) • (2π radians / 91.75766078 mi)**}** = 0.0513569 rad/min. One can combine the entire process by a single long expression as **{**(45 mi / hour) • (1 hour / 60 min) • (2π rad / 91.75766078 mi)**}** ≈ 0.05 rad / min. The units also properly cancel out except the radians and minutes, which is what was asked in the problem.

Answer: D

100. The rate of change of the measure of the central angle of the bicycle wheel is 12.5 radians per second. If the area of the bicycle wheel is 12 square feet, find the speed of the bicyclist (in miles per hour, to the nearest 0.1).

A. 16.7 mi/h

B. 8.3 mi/h

C. 0.3 mi/h

D. 33.3 mi/h

Using the given area of the bicycle wheel, we see that its radius must be 1.954410048 feet. The circumference of the bicycle wheel is then $2\pi(1.954410048) = 12.2799205$ feet. Again, as in the previous problem, we know that this distance is covered with an accompanying angle of 2π radians satisfied. So, we have $\frac{12.2799205}{2\pi} = 1.954410048$ ft/rad as the rate of change. Since there are 3,600 seconds in 1 hour, we have 12.5(3600) = 45,000 radians per hour covered at the constant bicycle speed. So we have 45,000(1.954410048) = 87,948.45216 feet per hour covered. Since 1 mile equals 5,280 feet, we have $87{,}948.45216(\frac{1}{5280}) = 16.65690382$ miles per hour as the speed of the bicycle. Again, this process can be expressed as **{**(12.5 rad / 1 s) • (3600 s / 1 h) • (12.27992069 ft / 2π rad) • (1 mi / 5,280 ft)**}** ≈ 16.7 mi/h. All the units except the miles and hours properly cancel out.

Answer: A

101. It costs 15 cents per cubic inch to build a bronze sphere-shaped statue with radius 5 feet. If the sculptor decides that the surface area of the statue should be changed so that it is three times the original, how much additional money (in dollars) is needed to build the new statue?

A. 705,205 dollars

B. 569,488 dollars

C. 135,717 dollars

D. 433,772 dollars

The radius of the statue is 5 feet, so that it is 5(12) = 60 inches. The volume of the statue is $(\frac{4\pi}{3})(60^3) = 904{,}778.6842$ cubic inches. Since it costs 15 cents per square inch to build the statue, the cost for this area will be 904,778.6842(15) = 13,571,680.26

cents. The surface area of the statue is $4\pi(60^2) = 45{,}238.93421$ square inches. Since the sculptor decides to increase this surface area by 3, the new surface area will be $3(45{,}238.93421) = 135{,}716.8026$ square inches. The radius of this larger statue will be: $(4\pi)r^2 = 135{,}716.8026$, so that the new radius $r = 103.9230484$ inches. The volume of the new statue is then $(\frac{4\pi}{3})(103.9230484^3) = 4{,}701{,}367.95$ cubic inches. The cost for this statue will be $4{,}701{,}367.95(15) = 70{,}520{,}519.26$ cents. Thus, the additional money in cents needed to build a new statue is $70{,}520{,}519.26 - 13{,}571{,}680.26 = 56{,}948{,}839$ cents. In dollars, this is about 569,488.

Answer: B

102. A rectangle encloses an ellipse. Its minor axis is equal to the half of the rectangle width (in meters), and major axis equals the rectangle length (in meters). Equation of the ellipse is given by $y = 6\sqrt{1 - x^2}$. Find the area of the rectangular region that does not include the enclosed ellipse (in square feet).

A. 24 ft^2

B. 96 ft^2

C. 29 ft^2

D. 314 ft^2

We start by drawing a picture from the information given. See the figure below.

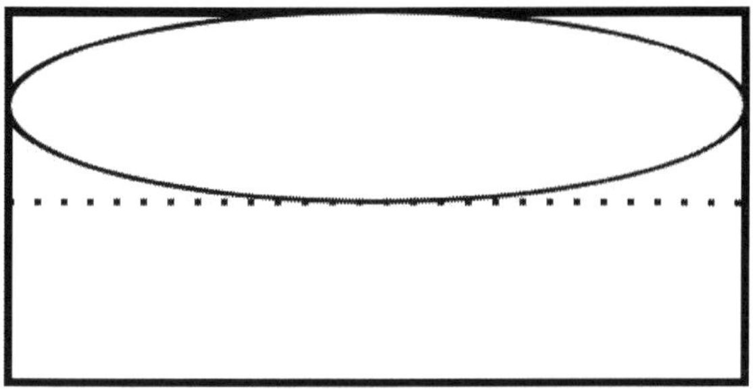

Note that we could have an ellipse right in the middle of the rectangle or on the bottom half – it will not change the area we need to find. From the ellipse equation given, namely $\boldsymbol{y} = \sqrt{\mathbf{1} - \boldsymbol{x}^2}$, we rewrite it (first squaring the *y* expression) and simplify it until we put it in standard form, which is $\left(\frac{y}{6}\right)^2 + \left(\frac{x}{1}\right)^2$ = 1. We now see that the major principal radius of the ellipse is 6, and minor principal radius is 1, which means that the two principal axes are twice these amounts, namely 12 and 2. Since the major axis equals the length of the rectangle, the length of the rectangle is 12. Minor axis equals half the rectangle width, so the rectangle width must be 2(2) = 4. Area of the rectangle is then 12(4) = 48. Area of the ellipse is 6 • 1 • π = 6π. Area of the desired rectangle region excluding the ellipse is then 48 – 6π = 29.15044408 square meters. In square feet this is 29.15044408(10.76391042) ≈ 314.

Answer: D

103. A circle with equation 73 = $(x + 6)^2 + (y - 4)^2$ encloses a concentric circle whose radius is a length from the origin to the center point (measured in centimeters). If the large circle is a wheel with rotation speed of 16 miles per hour, find the rotation speed (in feet per second) of the enclosed circle.

A. 71,301 ft/s

B. 23 ft/s

C. 20 ft/s

D. 84,480 ft/s

The equation of the large circle **73 = $(x + 6)^2 + (y - 4)^2$** means that the center of the circle is (-6, 4), which also means that the enclosed concentric circle has the same center point (-6, 4). Radius of the <u>large circle</u> is $\sqrt{73}$, so that its circumference is 2π($\sqrt{73}$) = 53.6835588 cm. To find the rotation speed of the enclosed circle, we need to find the number of revolutions the circle makes as it rotates. We find this using the circumference of the large circle we found (knowing that 1 revolution equals 53.6835588 cm), and we can put together all conversions in one long expression as follows: **{**(16 mi/h) • (5,280 ft/mi) • (0.3048 m/ft) • (100 cm/m) • (1 rev/53.6835588 cm)**}** = 47,965.34465 revolutions per hour.

Keep in mind that the number of revolutions is the same for both the large circle and enclosed smaller circle. Now, the radius of the enclosed circle is given as the distance from the origin to the center point, namely $\sqrt{6^2 + 4^2}$ = $\sqrt{52}$, so that its circumference is $2\pi(\sqrt{52}) = 45.3086936$ cm. We can now find the rotation speed of the smaller circle (in feet per second) by combining all necessary conversions again into one single expression: **{**(47,965.34465 rev/h) • (h / 3600s) • (45.3086936 cm/rev) • (1 in / 2.54 cm) • (1 ft / 12 in)**}** ≈ 20 ft/s. It is also important to know that the speed of rotation for the larger circle will be greater than the result we just found. We leave it to the reader to confirm this.

Answer: C

104. If a bicycle wheel whose equation is $y = \dfrac{43 - x^2 - 2x}{y - 4}$ (its radius measured in centimeters) rotates so that its rotation speed is 17 feet per second, what number of revolutions per minute must the wheel make to maintain this constant rotation speed?

A. 103

B. 2

C. 12

D. 714

The equation $y = \dfrac{43 - x^2 - 2x}{y - 4}$ can be rewritten as $y^2 - 4y + x^2 + 2x = 43$. To transform it into a standard equation of the circle, we need to complete the square: $x^2 + 2x + 1 - 1 + y^2 - 4y + 4 - 4 = 43$, which makes it $(x + 1)^2 + (y - 2)^2 = 48$. The radius of the bicycle wheel is then $\sqrt{48}$, so that its circumference is $2\pi\sqrt{48} = 43.53118474$ cm. Since the bicycle wheel rotation speed of 17 feet per second is given, we can find the number of revolutions per minute the bicycle wheel must make to maintain this speed: **{**(17 ft/s) • (60 s/min) • (2.54 cm/in) • (12 in/ft) • (1 rev / 43.53118474 cm)**}** ≈ 714 rev/min.

Answer: D

105. If a train wheel makes 20,000 revolutions per minute (its radius 0.5 meters), and the train begins to slow down at a steady rate of 50 ft/s each second, what

distance (in kilometers) does the train cover in 1 minute from the moment it begins to reduce its speed?

A. 436 km

B. 35 km

C. 3,386 km

D. 114,641 km

Even though this problem might look like a typical Physics problem (indeed it is usually solved using the formulas found in the subject), it can also be approximated using Arithmetic Sequence techniques. First, we will convert the number of revolutions per minute the train wheel makes to speed showing feet per second. To do this, we first need the circumference of the wheel, which is $2\pi(0.5) = \pi$ meters. The current speed of the train (in meters per minute) is then $20{,}000(\pi) = 62{,}831.85307$ m/min. The speed (in feet per second) of the train is calculated using conversions all at once: **{**(62,831.85307 m/min) • (100 cm/m) • (1 in / 2.54 cm) • (1 ft / 12 in) • (1 min / 60 s)**}** = 3,435.687504 ft/s.

Now, since the distance covered by the train decreases by 50 feet each second from the moment it began decreasing its speed, it covers 3,435.687504 – 50 = 3,385.687504 feet by the first second from the moment it began slowing down at a steady rate of 50 ft/s each second. We can think of the first term of an arithmetic sequence as d_1 = 3,385.687504 (the distance the train covers by the first second after it began slowing down). The 60th second will determine the distance the train covered from the 59th to the 60th second, and this is found as $d_{60} = d_1 + (n - 1)(r)$, where n is the last second in question (60th second in this case), and $r = -50$ (the sequence is decreasing by 50), so that d_{60} = 3,385.687504 + (60 – 1)(-50) = 435.687504. This is the distance the train covers during the 59th and 60th second interval (final second). Thus, the total distance traveled by the train in one minute (60 seconds) is $D = \frac{60(3{,}385.687504 + 435.687504)}{2}$ = 114,641.2502 feet. Converting this to kilometers, we get **{**(114,641.2502 ft) • (0.3048 m/ft) • (1 km / 1,000 m)**}** ≈ 35 km.

Answer: B

106. The itinerary of the two different car drivers is such that they both travel in straight line paths. Each path consists of a single lane. Car 1 path is the line $y = 2(x + 4) - 3$ (from left to right), and car 2 path is the line $y = 3(x - 12) + 2$ (from left

to right). Units are measured in meters. If both cars start from the *x*-axis, and car 1 travels at 65 meters per second, how fast does car 2 have to go (in miles per hour) for the drivers to meet each other?

A. 137 mi/h

B. 61 mi/h

C. 145 mi/h

D. 65 mi/h

Since the two cars start from the *x*-axis (where $y = 0$), we set each car path (line) to zero. We do this so that we know the *x*-value for each path where $y = 0$. For car 1, we have $0 = 2(x + 4) - 3$, so that $x = -\frac{5}{2}$. The starting point (coordinate) for car 1 will then be $(-\frac{5}{2}, 0)$. For car 2, we have $0 = 3(x - 12) + 2$, so that $x = \frac{34}{3}$. The starting point for car 2 is then $(\frac{34}{3}, 0)$. Now we have to find out where the two lines (car paths) meet in order to determine where the two drivers meet. This is done by setting the two given *y*--values (equations) equal: $2(x + 4) - 3 = 3(x - 12) + 2$, and solving for *x* we get $x = 39$. Plugging this *x*-value in either one of the two given equations, we get the *y*-value of 83. The meeting point for the two drivers is then (39,83).

Now, we must calculate the distance traveled for each of the two cars from their starting point to the point of their meeting. For car 1, the distance is $\sqrt{\left(39 + \frac{5}{2}\right)^2 + 83^2} =$ 92.79682107 meters. For car 2, the distance traveled to the meeting point is $\sqrt{\left(39 - \frac{34}{3}\right)^2 + 83^2}$ = 87.48968193 meters. The speed is given for the first car as 65 meters per second, which means that the time traveled for car 1 to the meeting point is 92.79682107 m (65 m/s) = 1.427643401 seconds. Car 2 must make the trip to the meeting point in the same amount of time in order for the drivers to meet. Thus, the speed for car 2 must be (87.48968193 m / 1.427643401 s) = 61.2825877 m/s. Converting to miles per hour we get (61.2825877 m/s) (3,600 s/h) (ft / 0.3048 m) (mi / 5,280 ft) ≈ 137 mi/h. Yes, they may collide at this incredible speed, but we want to think that they would not!

Answer: A

107. A giant scalene right triangle (viewed from the space shuttle) has a hypotenuse that is 100,000 feet in length. If bus *A* travels a total distance equal to the hypotenuse at 50 feet per second, and bus *B* travels the triangle base distance at 45 feet per second in the same time period as bus *A*, find the distance (in miles) of the height of the triangle that is perpendicular to the hypotenuse.

A. 43,589 miles

B. 4 miles

C. 7 miles

D. 8 miles

The triangle is shown in the graph below. The dotted line is the height extending from the hypotenuse, and this is what we need to find – in miles.

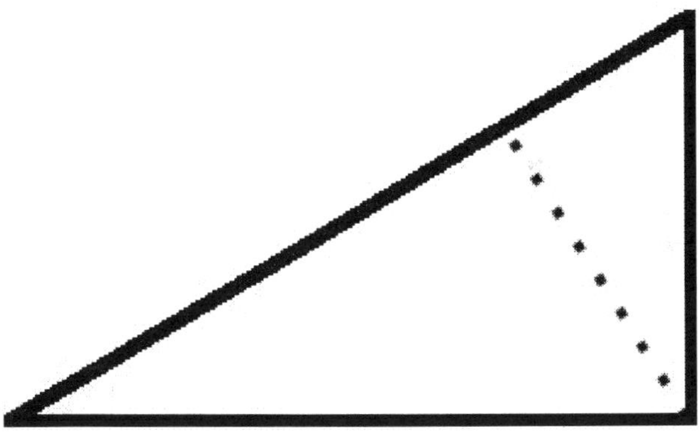

Since the hypotenuse is 100,000 feet and bus A travels at 50 ft/s, it travels the hypotenuse distance in (100,000 ft) / (50 ft/s) = 2,000 seconds. Bus B must travel the base distance of the triangle in 2,000 seconds as well, which means that the base distance traveled for bus B is (2,000 s) • (45 ft/s) = 90,000 feet. This is the distance of the base side of the triangle. The height of the triangle (its vertical edge) *H* is then $\sqrt{100{,}000^2 - 90{,}000^2}$ = 43,588.98944 feet. As you can see, the dotted height *h* in the figure separates the hypotenuse into two unequal parts; label the smaller part *x* and the larger part 100,000 – *x*. The dotted height *h* can be expressed using the two right triangles that the dotted height created. Thus the dotted height h is expressed as $h^2 = 90{,}000^2 - (100{,}000 - x)^2$, and $h^2 = 43{,}588.98944^2 - x^2$. Setting the two expressions for h^2 equal, we get $90{,}000^2 - (100{,}000 - x)^2 = 43{,}588.98944^2 - x^2$. Simplifying and solving for x we get $200{,}000x - 3{,}800{,}000{,}000 = 0$, so that $x = 19{,}000$. Plugging this x-value into either of the two equations for h^2 we get $h = 39{,}230.0905$ feet. In miles, this is 39,230.0905 ft (1 mi / 5,280 ft) ≈ 7 miles.

Answer: C

108. Suppose that a motorist travels on a circular road at 50 feet per second. He makes the round trip in 3 hours. If motorist *2* decides to go in a straight path along the diameter of this circular road, and decides to complete the round trip in the same time frame as the first motorist, but takes a 45-minute break at the semi-circle point before going back to the original point, what speed (not average), in kilometers per hour, is needed for motorist *2* to make the round trip in the same exact time period as the first motorist?

A. 23 km/h

B. 35 km/h

C. 27 km/h

D. 47 km/h

First we have to find the length of the circular road (its circumference) in kilometers. We do this using conversions: 50 ft/s • (0.3048 m/ft) • (km / 1,000 m) • (3,600 s/h) • (3h) = 164.592 km. Diameter d is found by $\pi d = 164.592$, so that $d = 52.39126079$ km. The three-hour trip means that 180 minutes (1 hour = 60 minutes) were spent on the trip for motorist 2. We are not asked to calculate his average speed. Since he took a 45-minute break, he was in motion for 180 – 45 = 135 minutes. Therefore we find his speed while keeping in mind that he travels two diameter distances of the circular road. Hence, we have his speed as **{**(2 • 52.39126079)km / 135 min**}** • (60 min/h) ≈ 47 km/h.

Answer: D

109. Car *A* travels on the line $y = 4x - 2$, at 55 feet per second and at a positive slope starting on the *x*-axis (from left to right). Units are measured in feet. Find an equation of the line with increasing slope (to the nearest 1 for slope and *y*-

intercept) of path for car *B* if it starts traveling at the same time as car *A* (also from the *x*-axis), with speed of 60 feet per second, and the two cars meet in exactly 1 minute.

A. $y = x + 1{,}644$

B. $y = 2x + 1{,}644$

C. $y = x + 1{,}316$

D. $y = 2x + 1{,}316$

Equation of a line $y = 4x - 2$ means that the *y*-axis point on the line (the *x*-value of starting point for car A) is $x = \frac{1}{2}$. We found this by setting $y = 0$. So the starting point for car A is $(\frac{1}{2}, 0)$. Now, the distance for car A according to the given speed is (55 ft/s)(60 s) = 3,300 feet. Using distance formula with the given point $(\frac{1}{2}, 0)$ we get $3{,}300 = \sqrt{\left(\frac{1}{2} - x\right)^2 + (0 - y)^2}$, so that $10{,}890{,}000 = \frac{1}{4} - x + x^2 + y^2$, so that $y^2 = 10{,}889{,}999.75 + x - x^2$. Using the first equation $y = 4x - 2$ and squaring it, we get $y^2 = 16x^2 - 16x + 4$, setting the two equations for y^2 equal we get $16x^2 - 16x + 4 = 10{,}889{,}999.75 + x - x^2$, which is simplified into $17x^2 - 17x - 10{,}889{,}995.75 = 0$. Solving this using the quadratic formula, we get $x = \frac{17 + 27{,}212.49713}{34} = 800.8675626$. We took a positive root because the graph is increasing to the right; therefore the *x*-value should be greater than $\frac{1}{2}$. Plugging this value into the equation of the line, we get $y = 4(800.8675626) - 2 = 3{,}201.47025$. Thus the coordinate of the meeting point for the two cars is (800.8675626, 3,201.47025).

Distance for car B from the starting point to the meeting point is (60 ft/s)(60 s) = 3,600 feet. Knowing that the starting point for car B will be the *x*-axis (where $y = 0$), and using the distance formula (as we did for car A) we get $12{,}960{,}000 = (800.8675626 - x)^2 + (3{,}201.47025 - 0)^2$, so that $x^2 - 1{,}601.735125x - 2{,}069{,}199.386 = 0$. Solving for *x* we get $x = \frac{1{,}601735125 - 3{,}292.772837}{2} = -845.5188559$. We took a negative root because we need an equation that is increasing to the right for car B, so that the *x*-value of the starting point for car B must be less than 800.8675626. The slope of the line is then $\frac{0 - 3{,}201.47025}{-845.5188559 - 800.8675626} = 1.944543647$. Equation of the line for car B is then $y = 1.944543647(x + 845.5188559)$, and rounding the slope and y-intercept to the nearest whole will give $y = 2x + 1{,}644$.

Answer: B

110. A cup is filled with water at 3 mL per second. If the cup's vertical height is 1 foot, base radiuses are 0.1 and 0.25 feet, find the time, in minutes, until the cup is $\frac{4}{5}$ full with water.

A. 14 min

B. 17 min

C. 16 min

D. 13 min

First we have to find the volume of the cup, and this is done by imagining a cone that will be cut into two pieces, so that one piece of it is the cup in question. Take a look at the figure below.

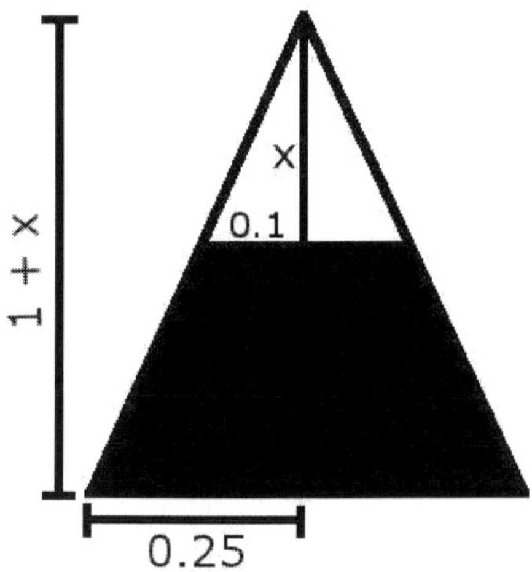

The cup is the shaded figure. As you can see, we have extended the cup into a cone with an additional piece for the height x. This height x is also the height for the small unshaded cone. It is clear that we can use direct proportion to find x. Since the height of the cup is 1, the height for the large composite cone is 1 + x. We then have $\frac{0.25}{1+x} = \frac{0.1}{x}$, which gives $x = \frac{2}{3}$. The height of the large composite cone is then $1 + \frac{2}{3} = \frac{5}{3}$. The volume of the cup is determined by subtracting the volume of the small cone from the volume of the large composite cone. The volume of the cup is then $(\frac{\pi}{9})(0.25^2(5) - 0.1^2(2))$ = 0.102101761 cubic feet. Since 1 mL = 1 cm^3, converting to cubic centimeters we get 0.102101761(28,316.84659) = 2,891.199903 cm^3 = 2,891.199903 mL. Since we need to find $\frac{4}{5}$ of this volume (water fills $\frac{4}{5}$ the volume of the cup), we need $(\frac{4}{5})$ • 2,891.199903 mL = 2,312.959922 mL. To find the time until this new volume is reached, we use unit conversions along with the given information: 2,312.959922 mL • (1 s / 3 mL) • (1 min / 60 s) ≈ 13 minutes.

Answer: D

111. Suppose that a cup with dimensions listed in question 110 was altered so that a closed right cylinder with radius 0.1 foot and height 1 foot was placed inside it, and water was filling the remaining volume of the cup at the same speed. What would be the time, in minutes, until water reaches the same height as in question 110?

A. 6 min

B. 7 min

C. 8 min

D. 9 min

First we find the height of the cup that corresponds to the $\frac{4}{5}$ of the volume of the cup that was found in question 110. We know that the height will be less than 1. We will deal with feet, as it will be easier. Again, we will add the small cone to the bottom portion of the cup so that we have a large composite cone. Since the volume of the cup in cubic feet was found as 0.102101761, $\frac{4}{5}$ of this volume is 0.081681409. This will create a height for the cup that will be less than 1. Adding the small cone whose volume was 0.006981317 cubic feet, the total volume (for new composite cone) will now be 0.081681409 + 0.006981317 = 0.088662726 cubic feet. Using the volume formula for the cone, we have 0.088662726 = $(\frac{\pi}{3})(r^2h)$. To avoid having two unknown variables, we can use direct proportion (having a composite cone and small cone) and solve for one unknown variable in terms of another unknown variable and then introduce it into the volume formula. Direct proportion gives us: $\frac{0.1}{\frac{2}{3}} = \frac{r}{h}$, so that $r = (\frac{3}{20})h$. Plugging this into the volume expression, we get 0.088662726 = $(\frac{\pi}{3})\{(\frac{3}{20})h\}^2h$, so that h = 1.555404369. This is the height of the composite cone, so that the height of the cup is 1.555404369 – $\frac{2}{3}$ (this is the height of the small cone) = 0.888737702 feet.

The volume of the cup when the poured water reaches this height is 0.081681409, as was found earlier. Now, to find the volume of the cup with water reaching this height after the cylinder with radius 0.1 foot is put in the cup is found by subtracting the volume of the cylinder up to this height from the overall volume of the cup (without the cylinder) up to this height. Thus have our volume V = 0.081681409 – $\pi(0.1^2)(0.888737702)$ = 0.053760891 cubic feet. To find the time until this volume is reached with the steady flow of water at the same speed as in question 110, we can convert all units in one expression as 0.053760891 ft^3 • (28,316.84659 cm^3 / ft^3) • (mL/cm^3) • (1 s / 3 mL) • (1 min / 60 s) ≈ 8 minutes. This result makes sense, because we significantly reduced the volume of the cup by putting in the closed cylinder, so the cup will indeed be filled faster.

Answer: C

112. In the following figure, *BD* and *AE* are medians. *EC* = 2*DC*. If area of triangle *ABC* is 100, find $_m$<*FDE*, $_m$<*FED* and area of triangle *FED*. (note: figure not drawn to scale).

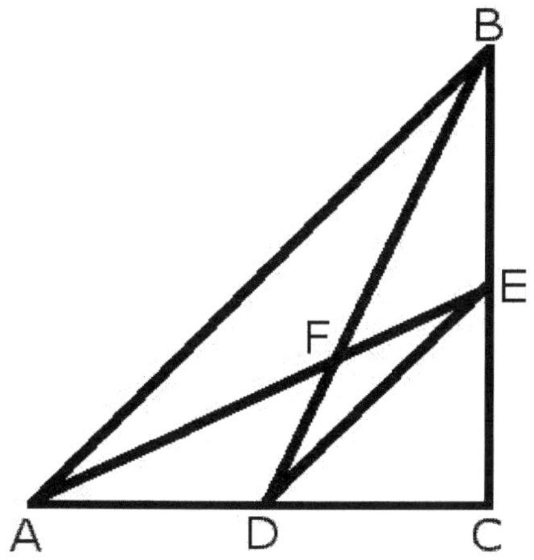

A. 17

B. 33

C. 38

D. 75

BD and AE are medians, which means that AD = DC, and BE = EC. Since the area of triangle ABC is 100, and knowing that EC = 2DC, we know that $100 = 2DC \cdot \frac{2(2DC)}{2}$, which means that DC = 5, EC = 10, BC = 20, and AC = 10. To find $_m$<AED, we see that $_m$<FED = $_m$<AEC – $_m$<DEC. $_m$<AEC = $\tan^{-1}(\frac{10}{10})$ = 45 degrees. $_m$<DEC = $\tan^{-1}(\frac{5}{10})$ = 26.56505118 degrees. Thus $_m$<FED = 45 – 26.56505118 = 18.43494882 degrees. $_m$<FDE = $_m$<FDC – $_m$<EDC. Since $_m$<FDC = $_m$<BDC, $_m$<BDC = $\tan^{-1}(\frac{20}{5})$ = 75.96375653 degrees, which is the measure of $_m$<FDC also. $_m$<EDC = $\tan^{-1}(\frac{10}{5})$ = 63.43494882 degrees. Thus $_m$<FDE = 75.96375653 – 63.43494882 = 12.52880771 degrees. Finding the area of triangle FDE requires separate work. Take a look at the figure below.

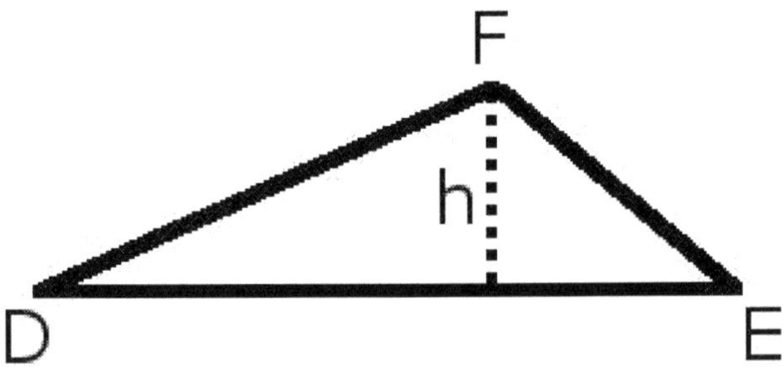

Since DE = $\sqrt{5^2 + 10^2}$ = 11.18033989, we can split DE into two horizontal lengths, *y* (horizontal distance covered by DF) and **{**11.18033989 – *y***}** (horizontal distance covered by FE). Note that we created two right angles split by the height *h*. Since we know the angles (<D and <E), we know that the left triangle has $_m$<F = 90 – 12.52880771 = 77.47119229 degrees. For the triangle on the right, $_m$<F = 90 – 18.43494882 = 71.56505118 degrees. We can use the *Law Of Sines* for each of the triangles: for the left triangle, we have $\frac{\sin((12.52880771)^o)}{h}$ = $\frac{\sin((77.47119229)^o)}{y}$, which means that *h* = 0.222222222*y*. For the triangle on the right, we have $\frac{\sin((18.43494882)^o)}{h} = \frac{\sin((71.56505118)^o)}{11.18033989 - y}$, which means that *h* = 3.726779963 – 0.333333333*y*. Setting the two equations for *h* equal, we have 0.222222222*y* = 3.726779963 – 0.333333333*y*, which gives *y* = 6.70820394, and *h* = 1.490711985. Thus, the area of triangle FDE is $\frac{11.18033989(1.490711985)}{2}$ ≈ 17.

Answer: A

113. An ellipse is perfectly enclosed in a trapezoid, so that its major axis matches the smaller base of the trapezoid, and its minor axis the trapezoid's height (see figure below). This graph exists in a standard *x-y* plane. The smaller base of the trapezoid is $\frac{2}{3}$ its larger base. Trapezoid height is $\frac{3}{5}$ its smaller base. The two slant heights of the trapezoid are congruent. If the leftmost vertex of the trapezoid has

coordinates (-5,6), and perimeter of the trapezoid is 190, find an equation of the ellipse and its area.

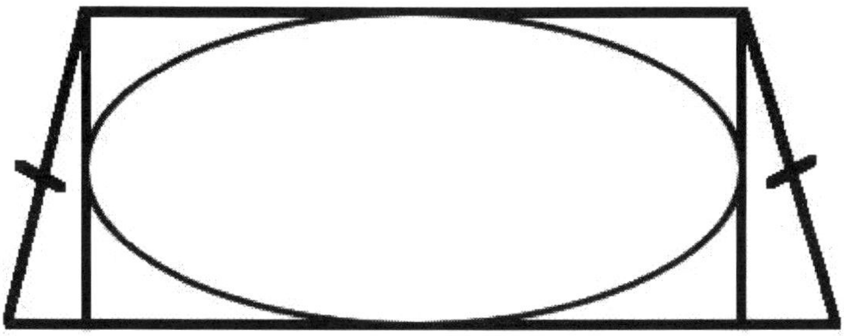

A. $\dfrac{(x-45)^2}{225}+\dfrac{(y-21)^2}{625}=1$, Area = 4,712

B. $\dfrac{(x-45)^2}{225}+\dfrac{(y-15)^2}{625}=1$, Area = 4,712

C. $\dfrac{\left(x-\frac{65}{2}\right)^2}{625}+\dfrac{(y-15)^2}{225}=1$, Area = 1,178

D. $\dfrac{\left(x-\frac{65}{2}\right)^2}{625}+\dfrac{(y-21)^2}{225}=1$, Area = 1,178

We can assign a dummy variable *x* to the large base, because everything starts from the large base. From the given information, we see that the large base of the trapezoid is $B = x$, smaller base $b = (\frac{2}{3})x$, height $h = (\frac{2}{3})(\frac{3}{5})x = (\frac{2}{5})x$. To find what *x* represents, we need to use the given trapezoid perimeter of 190. Before we do that, however, we must find the value of the slant height in terms of *x* (we need it for perimeter).

As you can see, the slant height acts as a hypotenuse of the right triangle in which the vertical height *h* participates. Since the base of this right triangle is $\dfrac{x-\left(\frac{2}{3}\right)x}{2} = (\frac{1}{6})x$ (why?), the slant height (hypotenuse of the right triangle) is $\sqrt{\left(\frac{2x}{5}\right)^2+\left(\frac{x}{6}\right)^2} = (\frac{13}{30})x$.

Thus the perimeter is 190 = $x + (\frac{2}{3})x + 2(\frac{13}{30})x$, so that x = 75. This is the length of the large base of the trapezoid. The smaller base $b = 75(\frac{2}{3}) = 50$ (this is also the major axis of the enclosed ellipse), and height $h = (\frac{2}{5})75 = 30$ (this is the minor axis of the enclosed ellipse).

Now, the large base B half is $\frac{75}{2}$ = 37.5. Since the coordinate of the leftmost trapezoid vertex is (-5, 6), this means that the x-value of the ellipse (and trapezoid) center is (-5 + 37.5) = 32.5 = $\frac{65}{2}$. The y-value of the center of the ellipse is (6 + 15) = 21. Thus the center of the ellipse (and trapezoid) has coordinates $(\frac{65}{2}, 21)$. Since the major axis of the ellipse corresponds to the length of the small base, that is, 50, the largest principal radius of the ellipse is $\frac{50}{2}$ = 25. The small principal radius of the ellipse is height h half = 15. Thus, the equation of the ellipse is $\frac{\left(x-\frac{65}{2}\right)^2}{625}+\frac{(y-21)^2}{225} = 1$. Area of the ellipse is $\pi(r_1)(r_2)$, where r_1 and r_2 are the principal radiuses of the ellipse. Thus the ellipse area is $\pi(25)(15) \approx 1{,}178$.

Answer: D

114. The following figure shows a circle enclosing an equilateral triangle that encloses a circle. If the small circle has radius that is equal to 5 feet, find the total area, in square centimeters, consisting of the three regions marked *y*.

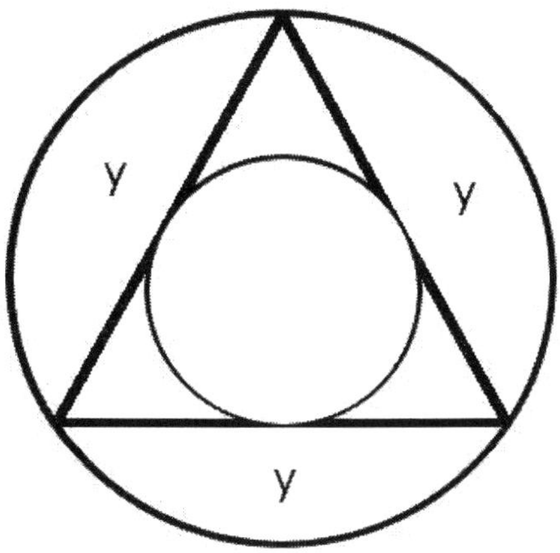

A. 130 cm ^{2}

B. 291,864 cm ^{2}

C. 184 cm ^{2}

D. 171,179 cm ^{2}

This problem is solved by subtracting the area of the equilateral triangle from the area of the large circle.

Using the figure above we can create two right triangles, and the variables to be used for them are r (small circle radius), R (large circle radius) and s (equilateral triangle side). The figure of the two right triangles below illustrates this.

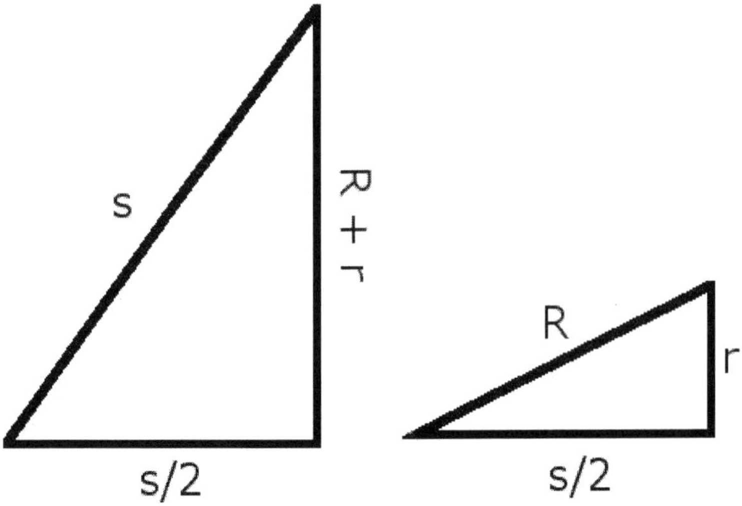

See if you can locate these right triangles on the original entire figure. It is easy to find them because one side is shared among the two triangles ($\frac{s}{2}$). Keep in mind that the two circles in the figure are concentric. Since one of the variables is known ($r = 5$), this simplifies our work in finding R and s.

Using the second triangle, we have $R^2 = 5^2 + \frac{s^2}{4}$, and using the first triangle we have $s^2 = (R + 5)^2 + \frac{s^2}{4}$. Rewriting and simplifying the second equation we have $(R + 5) = \frac{s\sqrt{3}}{2}$, so that $R = \frac{s\sqrt{3}}{2} - 5$. Plugging this R-value into the first equation we have $(\frac{s\sqrt{3}}{2} - 5)^2 = 25 + \frac{s^2}{4}$, and upon simplifying this leads to $s^2(\frac{3}{4}) - 5s\sqrt{3} + 25 = 25 + \frac{s^2}{4}$, which further leads to $\frac{s^2}{2} - 5s\sqrt{3} = 0$. Even greater simplification produces $s^2 - 10s\sqrt{3} = 0$. Upon factoring this expression we get $s(s - 10\sqrt{3}) = 0$, and we take the non-negative root $s = 10\sqrt{3}$, which makes $R = (10\sqrt{3})(\frac{\sqrt{3}}{2}) - 5 = 10$.

Area of the equilateral triangle is then $(10\sqrt{3})^2(\frac{\sqrt{3}}{4})$ = 129.9038106. Area of the large circle is $\pi(10^2)$ = 314.1592654. Thus the area of the large circle region that does not include the triangle is 314.1592654 – 129.9038106 = 184.2554548 square feet. Converting to square centimeters we get 184.2554548 ft^2 (144 in^2 / ft^2) • (6.4516 cm^2 / in^2) ≈ 171,179 cm^2.

Answer: D

115. The following figure shows a regular pentagon enclosing a kite whose three vertices divide the pentagon sides in half. If each side of the pentagon measures 10, find the area of the kite.

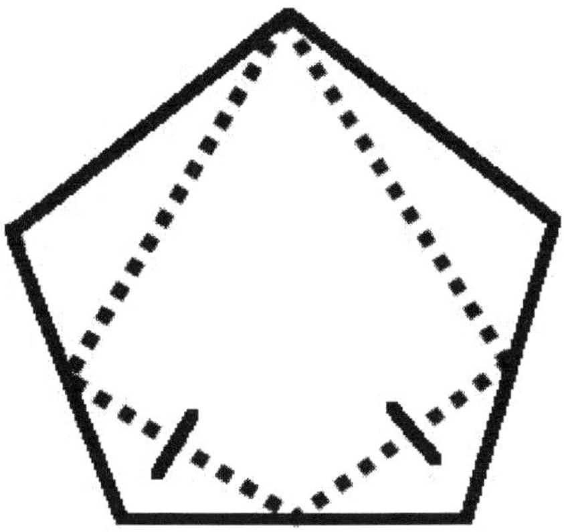

A. 31

B. 101

C. 70

D. 38

We can find the area of the kite by splitting it into two isosceles triangles and then simply finding the sum of their areas. We can also find the result using one height (combining two heights of the two parts of the kite) and base to find its area.

Interior angle measure sum for the pentagon is 180(5 – 2) = 540°, with each interior angle measuring $\frac{540}{5}$ = 108 degrees.

Let's do the problem by splitting the figure first. Since we are splitting the kite into two isosceles triangles, we will start with the upper triangle (larger part of the kite) with height h_1. The upper isosceles triangle is made up of two right triangles, so we will work with one of them. The hypotenuse for this triangle is the opposite side of the 108-degree pentagon angle, and we know the other two sides – they are $\frac{10}{2}$ and 10. We use the *Law of Cosines* to find the hypotenuse: $hyp_1{}^2 = 5^2 + 10^2 - 2(5)(10)\{\cos(108°)\}$, so that hyp_1 = 12.4860602. Now we find a right triangle we can work with to find the height h_1 for the upper isosceles triangle of the kite. Before we do that, we need an angle we can use with h_1 and base b, so we take a look at the top kite angle: since the two exterior angles (outside the kite) surrounding the top kite angle are equal (why?), we can find one of them and subtract the double amount of this angle measure from 108 to find the top kite angle measure. To find the measure of one of these exterior angles, label it C, we use the *Law of Cosines* again (the opposite side of this angle will be 5, half of pentagon side): $5^2 = 12.4860602^2 + 10^2 - 2(12.4860602)(10)\{\cos(C)\}$, so that C = $\cos^{-1}(0.924637939)$ = 22.38617758 degrees, so that the top angle of the kite is 108 – 2(22.38617758) = 63.22764484 degrees. Since we are splitting the top isosceles triangle (part of the kite) in half (we need to work with a right triangle to find the height h_1), the angle measure becomes $\frac{63.22764484}{2}$ = 31.61382242 degrees. We have $\cos(31.61382242°) = \frac{h_1}{12.4860602}$, so that h_1 = 10.6331351. Base b of the right triangle is also found: $\sqrt{(12.4860602)^2 - (10.6331351)^2}$ = 6.545084966, which makes large base B of the isosceles triangle 6.545084966(2) = 13.09016993. Thus the area of the upper isosceles triangle of the kite is $\frac{B(h_1)}{2} = \frac{13.09016993(10.6331351)}{2}$ = 69.59477268.

Area of the lower isosceles triangle (lower part of the kite) is found by first understanding the hypotenuse for one of the right triangles that make up the isosceles triangle. This hypotenuse is the opposite side of the 108-degree angle of the pentagon. If we split the hypotenuse in half, we get two right triangles with hypotenuse 5, and angle of measure 36 degrees (180° – 108° – 90°). Hypotenuse split in half produces two small bases. One of these bases is then: $\cos(36°) = \frac{base}{5}$, making base = 4.045084972, which makes the original hypotenuse equal to 2(4.045084972) =

8.090169944. Height h_2 for the bottom isosceles triangle is then easily found using the now known hypotenuse and half of base B (b) that we found earlier: $h_2 = \sqrt{(8.090169944)^2 - (6.545084966)^2} = 4.75528259$. Thus the area of the bottom isosceles triangle is $\frac{4.75528259(13.09016993)}{2} = 31.12372858$. Adding the two areas together we get the area of the kite as $31.12372858 + 69.59477268 \approx 101$.

Now, the quicker way to solve this would be to combine the two heights, h_1 and h_2, and use the base B. The area is then simply $\frac{(10.6331351+4.75528259)(13.09016993)}{2}$ ≈ 101.

Answer: B

116. The following figure shows a slice of pizza with a large circular pepperoni piece that is concentric with a larger circle whose visible part is the slice's circular edge. A vertical chord is tangent to the small circle and makes an isosceles triangle with two identical sides that are also tangent to the small circle. If the small circle radius is $\frac{3}{5}$ the radius of the large circle, find the ratio, to the nearest hundredth, of the area of the pepperoni piece to the area of the slice.

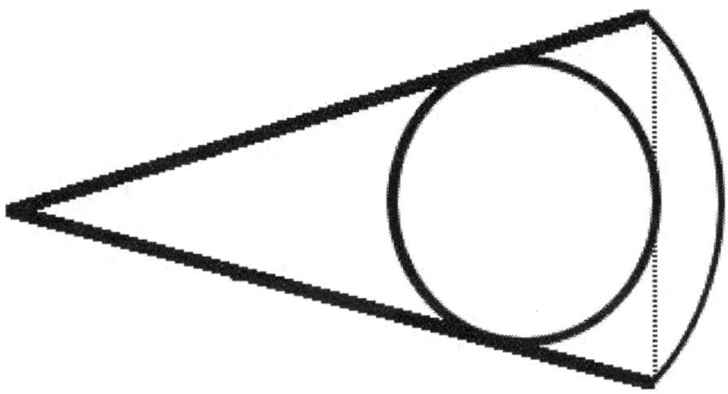

A. 0.43

B. 0.36

C. 0.52

D. 0.83

This problem involves working with addition and subtraction of areas of certain regions in the figure before arriving at the desired answer. See the figure below.

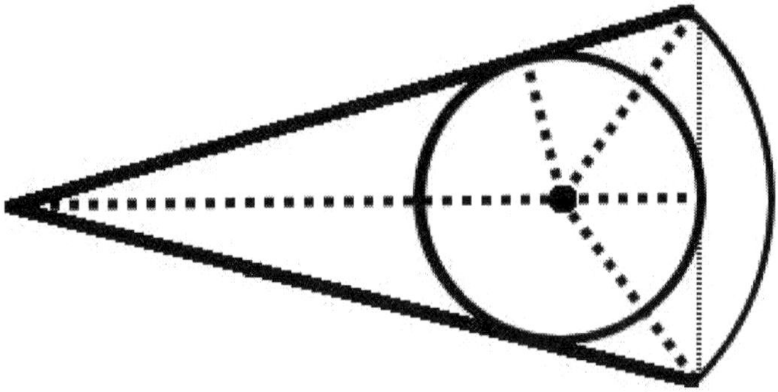

If we let small radius $r = 1$, large radius $R = \frac{5}{3}$. We will start with an isosceles triangle that has a height equal to the small radius 1 (two equal sides are equal to the large radius $\frac{5}{3}$, and vertex is the center of the small circle). Splitting it into two right triangles, we see that we have dimensions the height, 1, and hypotenuse, $\frac{5}{3}$. The base is then $\sqrt{\left(\frac{5}{3}\right)^2 - \left(\frac{3}{3}\right)^2} = \frac{4}{3}$. This makes the base for the isosceles triangle (and the parent isosceles triangle that almost takes an entire figure) equal to $2(\frac{4}{3}) = \frac{8}{3}$. Also, the angle between the height and hypotenuse for each of the two small right triangles measures $\cos^{-1}\{\frac{1}{\frac{5}{3}}\} = 53.13010235$ degrees. The area of the isosceles triangle (drawn from the center of the circle) is then $(\frac{8}{3}) \cdot \frac{1}{2} = \frac{4}{3}$. Area of the small circle is $\pi(1^2) = \pi$.

We now need to find the height H for the parent isosceles triangle (the largest isosceles triangle in the figure), which consists the small circle diameter (2) and additional segment outside the circle that connects to the vertex of the triangle. We will name this segment x. We can work with a right triangle with dimensions that are hypotenuse, $1 + x$, and height 1 (this triangle has a vertex that is the center of the small circle). To find x, we can use the angle between the height of 1 and hypotenuse $1 + x$. We will name this angle A. Since the adjacent angle is equal in measure to the angle we found earlier (why?), we can subtract the angle measure that is twice the angle measure we found earlier from 180 degrees. Take a look at the kite-like figure and try to see why the two angles are equal in measure. So we have $A = 180^\circ - 2(53.13010235^\circ) = 73.73979529$ degrees. Now we have $\cos(73.73979529^\circ) = \frac{1}{1+x}$, which makes $x = 2.571428571$. This makes the parent isosceles triangle height $H = 2 + 2.571428571 = 4.571428571$. The area of the parent isosceles triangle A_T is then $(\frac{8}{3}) \cdot 4.571428571 / 2 =$ 6.095238096.

We now have to calculate a part of the area, label it A_c, taken by the large circle (whose radius is $\frac{5}{3}$), and this is done by direct proportion, knowing that the central angle (this is the largest angle of the isosceles triangle we worked with earlier with vertex the center of the small circle), is $2(53.13010235^\circ) = 106.2602047$ degrees. The ratio of this angle to 360 degrees is $\frac{106.2602047}{360} = 0.295167235$, which means that the area that the part of the large circle takes as defined by the central angle will also take 0.295167235 of its complete area, that is, $A_c = \pi(\frac{5}{3})^2 \cdot 0.295167235 = 2.57582005$. We now have A_T (area of the largest isosceles triangle), A_t (area of the small isosceles triangle), and A_c (area of the part of the large circle). The desired area of the pizza slice is then calculated as $A_s = A_T + A_c - A_t$. We subtract the area of the small isosceles triangle because we have double counted it with A_c. We have $A_s = 6.095238096 + 2.57582005 - \frac{4}{3} = 7.337724813$. Since the area of the small circle is π, the ratio of the area of the small circle (pepperoni piece) to the area of the pizza slice is $\frac{\pi}{7.337724813} \approx 0.43$.

Answer: A

117. If it costs 2 dollars per inch to build a figure, how much money will it cost to build the following 2-dimensional figure consisting of a semi-circle (its radius is given in inches) joined by an oblique triangle which is formed by cutting out a right triangle below it?

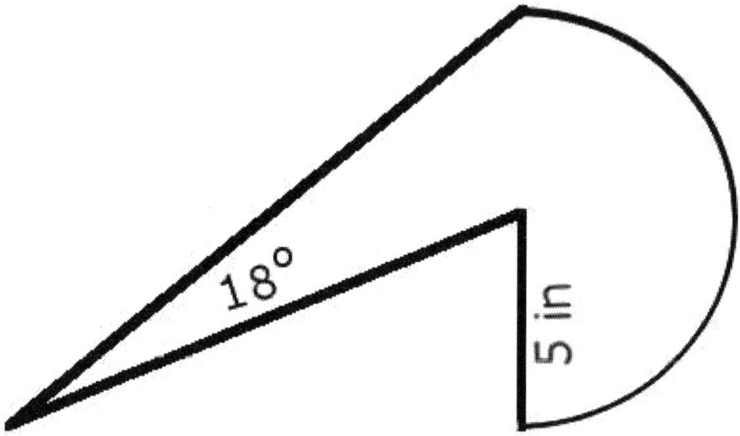

A. $157

B. $215

C. $107

D. $264

This problem is solved by using trigonometric angle addition for tangent. We can add the base to the figure to see how this will work.

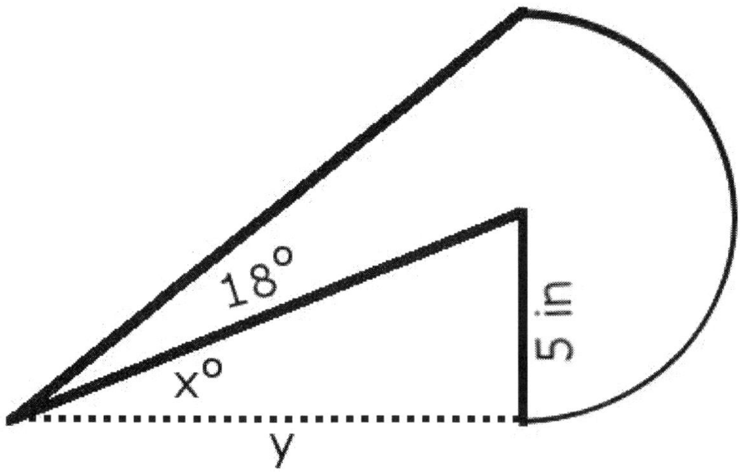

We see that there are two right triangles associated with this figure that we can work with. The larger right triangle of the two will have base *y*, height 5(2) = 10, and angle measure of 18 + *x*. The smaller of the two will have the same base *y*, angle *x*, and height 5. Thus $\tan(18 + x) = \frac{10}{y}$, and $\tan(x) = \frac{5}{y}$. Using trigonometric angle addition formula we have $\tan(18 + x) = \frac{\tan(18^o) + \tan(x)}{1 - (\tan(18^o))(\tan(x))} = \frac{\tan(18^o) + \frac{5}{y}}{1 - \frac{(\tan(18^o))5}{y}}$. Since $\tan(18 + x) = \frac{10}{y}$, $\frac{\tan(18^o) + \frac{5}{y}}{1 - \frac{(\tan(18^o))5}{y}} = \frac{10}{y}$, which simplifies algebraically to $\tan(18^\circ) - \frac{5}{y} + \frac{50}{y^2\tan(18^o)} = 0$. Putting this result in standard quadratic form we have $y^2(\tan(18^\circ)) - 5y + 50\tan(18^\circ) = 0$. Using quadratic formula we have $y = \frac{5 + 1.971151491}{2\tan(18^o)} = 10.72749909$. We reject the second root because it produces an answer that does not agree with the circle dimensions.

Thus the area of the non-right triangle is calculated as the difference between the areas of the small and large right triangles, that is, $(10.72749909 \bullet \frac{10}{2}) - (10.72749909 \bullet \frac{5}{2}) = 26.81874773$. Area of the semi-circle is $\frac{\pi(5^2)}{2} = 39.26990817$. Area of the entire figure is then 39.26990817 + 26.81874773 = 66.0886559 square inches. Since it costs 2 dollars to build an inch of the figure, it will cost $2^2 = 4$ dollars to build an area measuring one square inch. Thus the total cost to build the figure in this example is 4(66.0886559) ≈ 264 dollars.

Answer: D

118. Find the area of a triangle formed by ellipse foci and ellipse point (7, 2.8) if the equation of the ellipse is $36x^2 - 288x + 25y^2 + 100y - 224 = 0$.

A. 16

B. 10

C. 4

D. 19

First we must put the ellipse equation in standard form. We do that by completing the square and simplifying. We have $36(x^2 - 8x + 16) - 576 + 25(y^2 + 4y + 4) - 100 - 224 = 0$, which simplifies to $36(x - 4)^2 + 25(y + 2)^2 = 900$. Making sure that we have 1 on the right side of the equation, we divide both sides by 900 and thus have $\frac{(x-4)^2}{25} + \frac{(y+2)^2}{36} = 1$. The center of the ellipse is then (4, -2), and vertices at (4, -2 + 6) and (4, -2 – 6). Note that the major axis is the vertical axis in this example.

Since $a = 6$ and $b = 5$, our foci will be $c^2 = a^2 - b^2 = 36 - 25 = 11$, so that $c = \pm\sqrt{11}$, which means that foci are $\pm\sqrt{11}$ away from the center of the ellipse and located on its major axis (vertical axis), and their coordinates are $(4, -2 - \sqrt{11})$ and $(4, -2 + \sqrt{11})$.

Now, since we are building a triangle with points $(4, -2 - \sqrt{11})$, $(4, -2 + \sqrt{11})$ and (7,2.8), we need to analyze what this triangle may be like. We will start by determining distances between the three points to know its sides. We have $d_1 = 2\sqrt{11} = 6.633249581$, $d_2 = \sqrt{(7-4)^2 + \left(2.8 + 2 + \sqrt{11}\right)^2} = 8.653299832$, $d_3 = \sqrt{(7-4)^2 + \left(2.8 + 2 - \sqrt{11}\right)^2} = 3.346700168$, so that the triangle is scalene.

Since the two points (foci) are both on the vertical axis (where $x = 4$), the height is determined from the vertical line connecting the foci to the third point (7, 2.8). It is clear that the triangle is a non-right triangle, because the y-values for coordinate (7, 2.8) is much bigger than the greatest y-value for one of the foci whose coordinates are $(4, -2 + \sqrt{11})$. This makes it obvious that the height must be drawn from the line segment connecting the two points (7, 2.8) and farthest focus $(4, -2 - \sqrt{11})$ to the third point which is the close focus $(4, -2 + \sqrt{11})$. We already know this line segment, which is d_2 that we found earlier. The height h, which is the shortest distance from this line segment to the third point that is the closest focus, is found through the line-to-point distance formula. Before we use it, however, we must find an equation of the line on which the line segment d_2 lies: since the slope is $\frac{-4.8 - \sqrt{11}}{-3} = 2.705541597$, $y = 2.705541597(x - 7) + 2.8 = 2.705541597x - 16.13879118$. Rewriting this equation we have $2.705541597x - y - 16.13879118 = 0$. Using the line-to-point distance formula, we have $h = \frac{|2.705541597(4) + (-1)(-2 + \sqrt{11}) + (-16.13879118)|}{\sqrt{(2.705541597)^2 + (-1)^2}} = 2.299671701$. Thus the area of the triangle is $\frac{2.299671701(8.653299832)}{2} \approx 10$.

Answer: B

119. The following figure shows a heel part. Its vertical height is 10. Find the area of this concave polygon.

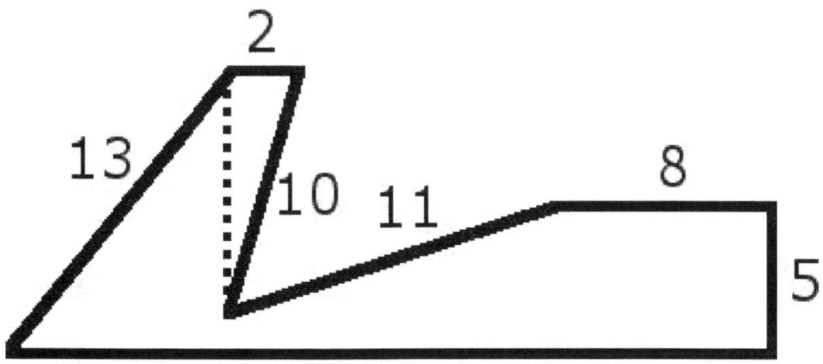

A. 106

B. 117

C. 116

D. 118

One must be careful in determining the area for each region before estimating the final area of the figure. Let us look at the figure again and take one region at a time (from left to right).

Since the vertical height is 10 (be sure that you see that the dashed height is not 10), the horizontal distance that the hypotenuse of 13 covers is $\sqrt{13^2 - 10^2}$ = 8.306623863. Thus the area of the first region $A_1 = \frac{8.306623863(10)}{2}$ = 41.53311931.

The second region has a vertical dashed line that is $\sqrt{10^2 - 2^2}$ = 9.797958971. Thus $A_2 = \frac{2(9.797958971)}{2}$ = 9.797958971.

The third region can be split into a rectangle and triangle. The rectangle consists of a vertical component we need to find and base (this base length or horizontal length is shared among the rectangle and triangle) that the hypotenuse of 11 covers. The small vertical component is the difference between the vertical height of the figure and the dashed height that we found for the second region. It is thus 10 – 9.797958971 = 0.202041029. The vertical length that the triangle with hypotenuse 11 covers is 5 – 0.202041029 = 4.797958971. The base length for both this rectangular region and the triangle is thus $\sqrt{11^2 - (4.797958971)^2}$ = 9.898464008. Thus the rectangle area A_3 = 9.898464008(0.202041029) = 1.999895854 and triangle area $A_4 = \frac{9.898464008(4.797958971)}{2}$ = 23.74621209. The last region is a rectangle whose area A_5 = 5(8) = 40. Thus the total area of the figure is 40 + 23.74621209 + 1.999895854 + 9.797958971 + 41.53311931 ≈ 117.

Answer: B

120. A line segment is defined by two endpoints whose coordinates are (-3, 17) and (6, -19). A third point with coordinates (8, -2) makes a triangle with the two points. Find the area of this triangle.

A. 44

B. 88

C. 545

D. 148

The equation of the line on which the segment consisting of two endpoints is drawn is $y = -4(x + 3) + 17 = -4x + 5$, where the slope $m = \frac{-19 - 17}{6 - (-3)} = -4$. If we rewrite the equation so that we have a zero on the right side of the equation we will have $4x + y - 5 = 0$. The shortest distance (perpendicular distance) from this line to the third point at (8, -2) is $\frac{|4^2(8) + 1^2(-2) + (-5)|}{\sqrt{4^2 + 1^2}} = 29.34681063$. This serves as a height for the triangle that is created with the third point. The base length of this triangle will be the line segment made by the first two given points, so we must find its distance using the distance formula: $d = \sqrt{(6 + 3)^2 + (-19 - 17)^2} = 37.10795063$. Thus the area of the triangle created by the three points is $\frac{37.10795063(29.34681063)}{2} \approx 545$.

Answer: C

121. A torus has an inner radius R that is $\frac{5}{3}$ the outer radius r. If the largest torus can be put in a right circular cylinder whose radius is 4 and height 7, what is the maximum volume of a smaller right cylinder, with the same height 7, which can be perfectly stacked along with the torus in the large cylinder?

A. $\frac{175\pi}{9}$

B. $\frac{63\pi}{4}$

C. 112π

D. 7π

Since the inner radius is $\frac{5}{3}$ the outer radius, this means that $R = (\frac{5}{3})r$. The radius of the large cylinder will equal the sum of the outer and inner radiuses, namely $r + (\frac{5}{3})r = (\frac{8}{3})r$. See the accompanying figure below for clarity.

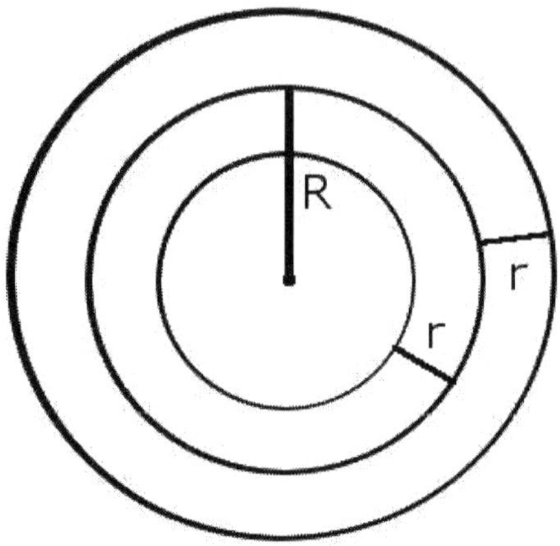

The vertical line segment drawn in the circle pertains to the length of the inner radius R, and the two small line segments drawn pertain to the length of the outer radius r. Since the torus takes a diameter equal to $2r$, the small cylinder radius that we must fit into the remaining circle (smallest circle in the figure) is $(\frac{8}{3})r - 2r = (\frac{2}{3})r$. In other words, the largest possible smaller cylinder will have a base circle that is the smallest circle in the figure, because the torus will take $2r$ in radius space from the initial $(\frac{8}{3})r$ radius space of the large cylinder. Since the initial large cylinder has a radius equal to 4, this means that $4 = (\frac{8}{3})r$, so that $r = \frac{3}{2}$. This makes the smaller cylinder radius be $(\frac{2}{3})(\frac{3}{2}) = 1$. Thus the volume of the smaller cylinder that will perfectly fit in the remaining circular space of the large cylinder after the torus is placed inside will be $\pi(1^2)(7) = 7\pi$.

Answer: D

122. The following figure shows a question mark consisting of a semi-torus, oblique cylinder, right cylinder, and ellipsoid. All figures have the same radius (indicated by horizontal lines where semi-torus' outer radius equals the radius of the other figures), and all are empty inside. Oblique and right cylinders have the same vertical height. If the largest circle diameter of the semi-torus' equals 100 ft, ellipsoid's smaller principal radiuses are both half the largest principal radius, and ratio of the oblique cylinder's surface area to its volume is $\frac{1}{9}$, find the volume, in cubic yards, of the entire figure.

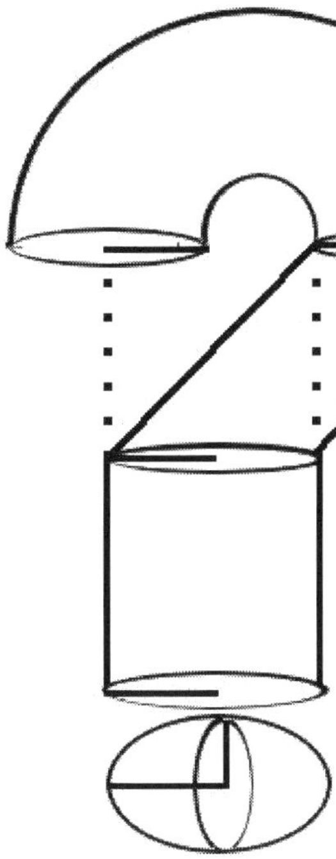

A. 155,613 yd ^{3}

B. 5,763 yd ^{3}

C. 10,150 yd ^{3}

D. 274,048 yd ^{3}

It is important to find the radius r of the figures, because it is the same for all of them. Semi-torus' outer radius is equal to this common radius r, and ellipsoid's principal radius (largest radius) is also equal to this common radius r. Let's take a look at the figure again.

Since the semi-torus' largest circle diameter is 100 (in feet), its radius must be 50. This radius will take quantity that is twice the radius that we need to find, plus the smallest circle radius which we can label as x. We now have $50 = 2r + x$. The inner radius of the semi-torus, $R = 50 - r$ (it starts from the center of the smallest circle and stops at the end of horizontal line whose length is r).

Using the dashed vertical lines that correspond to the right cylinder dimensions, namely r, we can see that the cylinder's diameter, namely $2r = 2x + r$ (this is what it means in the semi-torus' figure). This makes $x = \frac{r}{2}$. Plugging this in the first equation we have $50 = 2r + \frac{r}{2} = \frac{5r}{2}$, so that $r = 20$. This makes $x = 10$, and $R = 50 - 20 = 30$. The volume of the semi-torus V_t is then $\frac{\pi(20^2)2\pi(30)}{2} = 118{,}435.2528$ cubic feet.

Now, the volume of the oblique and right cylinders is a little bit trickier to find. The vertical height is the same for both, but it is not given. We must find it using the fact that the ratio of the oblique cylinder surface area to its volume is $\frac{1}{9}$, so that $\frac{1}{9} = \frac{2\pi r h_s}{\pi r^2 h_s \sin(a)} = \frac{2}{r\sin(a)}$, where a is the angle between the oblique cylinder slant height h_s and circular base. Since we already know $r = 20$, we now have $\frac{1}{9} = \frac{1}{\mathbf{10}\sin(a)}$, so that $a = \sin^{-1}(\frac{9}{10}) = 64.15806724$ degrees. Using the dashed vertical height in the figure that equals the vertical of the right cylinder, we can now build a right triangle to find the vertical height for both oblique and right cylinders:

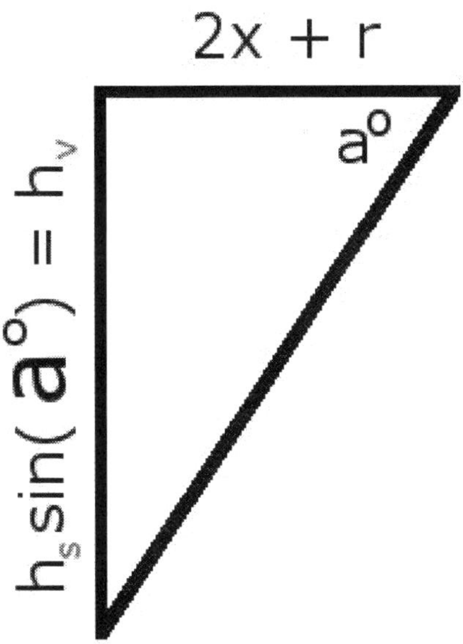

Since $2x + r = 2r = 2(20) = 40$, we find the vertical height h_v using the usual trigonometry: $\sin(64.15806724°) = \frac{h_v}{40}$, which gives the vertical height $h_v = 40(\frac{9}{10}) = 36$. Thus the combined volume V_{o+r} of the right and oblique cylinders is $2(20^2 \cdot 36) =$ 28,800 cubic feet.

Since the second and third principal radiuses r_2 and r_3 of the ellipsoid are both half the first principal radius that is $r = 20$, so that $r_2 = r_3 = 10$, the volume of the ellipsoid V_e is $(\frac{4\pi}{3})(20)(10^2) = 8{,}377.58041$ cubic feet. Combining all the volumes together we have the volume of the entire figure $V = 8{,}377.58041 + 28{,}800 + 118{,}435.2528 =$ 155,612.8332 cubic feet. In cubic yards this is $155{,}612.8332(0.037037037) \approx 5{,}763$.

Answer: B

123. The following non-right quadrilateral *ABCD* shows four distinct sides and diagonals *DB* and *AC*. If *AF* = *DF*, *AC* = 5, and *DC* = 3.5, find the area of this quadrilateral. **(note: figure not drawn to scale)**

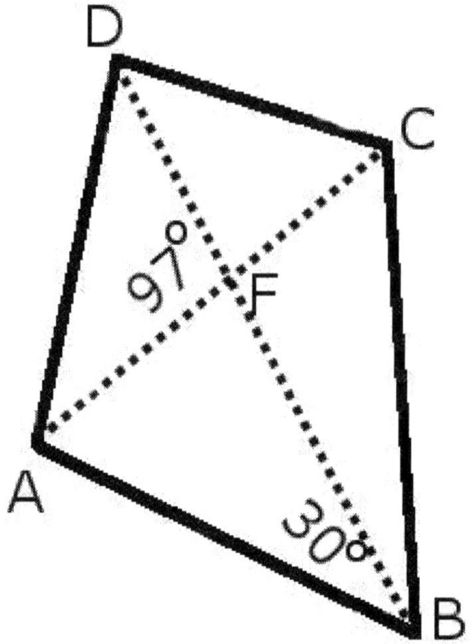

A. 23

B. 55

C. 33

D. cannot be determined

This problem requires one to find all four sides and measures of two opposite angles, namely $_m$<DAB and $_m$<DCB, to find the area. Let's look at the figure again.

We are given diagonal AC = 5, DC = 3.5 and AF = DF. Since the central angle of triangle AFD is 97 degrees, and AF = DF, $_m$<DAF = $_m$<ADF = $\frac{180-97}{2}$ = 41.5 degrees. Since $_m$<ADF = 41.5, $_m$<ADC = $(41.5 + x)^{\circ}$, where x = $_m$<FDC. Using the *Law of Sines* for triangle ADC, we have $\frac{\sin(41.5^{o})}{3.5} = \frac{\sin((41.5+x)^{o})}{5}$, so that $_m$<ADC = 71.19125623 degrees, and x = 29.69125623 degrees. This makes $_m$<DCF = 180 – 83 – 29.69125623 = 67.30874377 degrees. Using the *Law of Sines* again for triangle ADC

we have $\dfrac{\sin(41.5^o)}{3.5} = \dfrac{\sin(67.30874377^o)}{AD}$, which makes AD = 4.873214155. We could have used the *Law of Cosines* initially to find AD with triangle ADC, but we would not have known which root to take after solving with quadratic formula, because the figure is not drawn to scale.

Using the *Law of Sines* for triangle ABD, we have $\dfrac{\sin(41.5^o)}{AB} = \dfrac{\sin(30^o)}{4.873214155}$, which gives AB = 6.458178797. Now, $_m$<FAB = 180 – 30 – 83 = 67 degrees, so that when using the *Law of Cosines* with triangle ABC we have BC = $\sqrt{5^2 + (6.458178797)^2 - 2(5)(6.458178797)\cos(67^o)}$ = 6.440027832. We now know all four sides of this quadrilateral and need to use the opposite angles <DAB and <DCB to find the area. We already know that $_m$<DAB = 41.5 + 67 = 108.5 degrees. To find $_m$<DCB, we will use the second diagonal DB using the *Law of Cosines* for both angles <DAB and <DCB, and set up two equations for DB:

$DB^2 = 4.873214155^2 + 6.458178797^2 - 2(4.873214155)(6.458178797)\cos(108.5^o)$,

$DB^2 = 3.5^2 + 6.440027832^2 - 2(6.440027832)(3.5)\cos({}_m\text{<DCB})$. Setting these two equations for DB equal, we have

$4.873214155^2 + 6.458178797^2 - 2(4.873214155)(6.458178797)\cos(108.5^o) = 3.5^2 + 6.440027832^2 - 2(6.440027832)(3.5)\cos({}_m\text{<DCB})$, and after simplifying we get $_m$<DCB = $\cos^{-1}(-0.703298013)$ = 134.6922066 degrees. Using the general formula to find the area of a quadrilateral, we have *A* =

$$\sqrt{(s-4.873214155)(s-3.5)(s-6.440027832)(s-6.458178797) - (4.873214155)(3.5)(6.440027832)(6.458178797)\left(\cos\left(\left(\frac{134.6922066+108.5}{2}\right)^o\right)\right)^2}$$

, where $s = \dfrac{4.873214155 + 3.5 + 6.440027832 + 6.458178797}{2}$ = 10.63571039.

Plugging this into the area formula we get A ≈ 23.

Answer: A

124. In the following figure of triangle ABC, if AB = 8, find the measure of angle *x*.

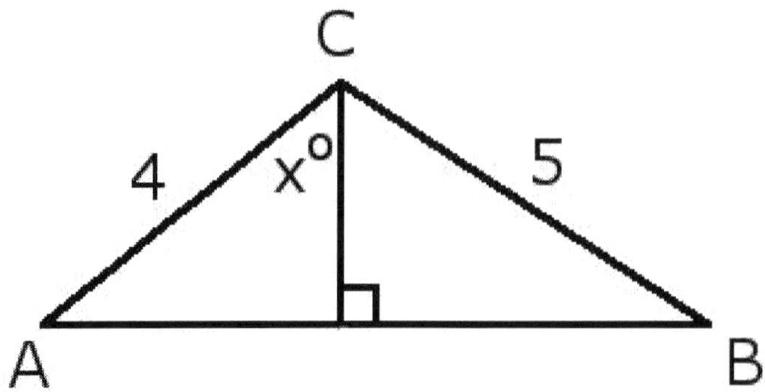

A. 59°

B. 57°

C. 55°

D. 61°

We first split the base, which is 8, into two bases for each of the two small triangles that make up the parent triangle. We can label the left triangle (which has angle *x*) base as *y*, and the other triangle base as 8 – *y*. Since both triangles share the height, the first triangle will have the height as $\sqrt{4^2 - y^2}$, and the second triangle will have the height as $\sqrt{5^2 - (8 - y)^2}$. Setting the two expressions for the height equal, and subsequently squaring and simplifying them, we have $16 - y^2 = -39 + 16y - y^2$. Simplifying further we have $16y - 55 = 0$. This makes $y = \frac{55}{16}$ (base length for the left right triangle). We do not need the other right triangle base, because we are asked to find the angle *x*, which is in the first right triangle. Angle *x* is then $\sin^{-1}(\frac{\frac{55}{16}}{4}) \approx 59$ degrees.

Answer: A

125. The center of the circle has coordinates (1, 4), and a point lying on the circle has coordinates (-3, -6). Find the equation of this circle.

A. $(x - 1)^2 + (y - 4)^2 = \sqrt{116}$

B. $(x + 3)^2 + (y + 6)^2 = 116$

C. $(x - 1)^2 + (y - 4)^2 = 116$

D. $(x + 3)^2 + (y + 6)^2 = \sqrt{116}$

We only need to find the radius of the circle to find the equation. Using the distance formula, we have radius $r = \sqrt{(-3-1)^2 + (-6-4)^2} = \sqrt{116}$. Since the equation of the circle written in standard form requires the radius to be squared, and the circle center is at (1, 4), the equation is thus $116 = (x - 1)^2 + (y - 4)^2$.

Answer: C

126. Given that an ellipse has vertices at (-4, 7), (-4, 3), (2, 5) and (-10, 5), find the coordinates of its foci.

A. $(4\sqrt{2}, 5)$, $(-4\sqrt{2}, 5)$

B. $(4\sqrt{2} - 4, 5)$, $(-4\sqrt{2} - 4, 5)$

C. $(4\sqrt{2} + 4, 5)$, $(-4\sqrt{2} + 4, 5)$

D. $(2 - 4\sqrt{2}, 5)$, $(-12 + 4\sqrt{2}, 5)$

This ellipse has a major axis parallel to the *x*-axis, and the center is at (-4, 5). The larger principal radius from the center is the horizontal radius *a*, which is 6. We know this from seeing the horizontal distance between center and one horizontal vertex, (-4, 5) and (2, 5). The second, smaller principal radius from the center is $b = 2$. We see this when comparing the vertical distance of the center and vertical vertex, (-4, 5) and (-4, 7). Foci, which are located a certain horizontal distance from the center in both directions, are calculated using the formula $\sqrt{a^2 - b^2} = \sqrt{36 - 4} = 4\sqrt{2}$. Using this distance from the center point in both directions, we see that the foci's coordinates are $(-4 - 4\sqrt{2}, 5)$ and $(-4 + 4\sqrt{2}, 5)$.

Answer: B

127. In the following figure of a right triangle, if *cos* ɵ = $\frac{\sqrt{2}}{2}$, find *r*.

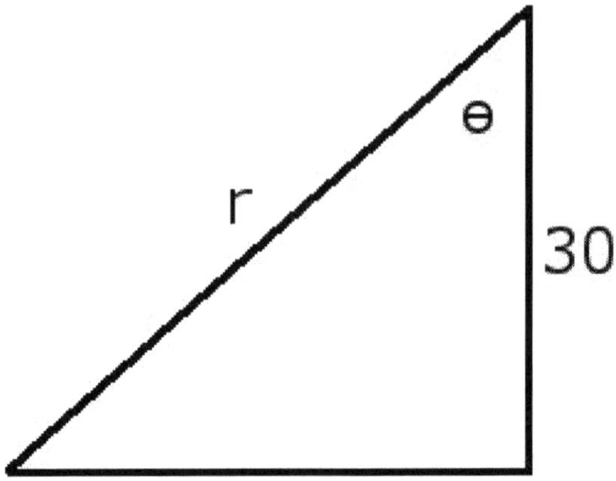

A. $30\sqrt{2}$

B. $60\sqrt{2}$

C. 35

D. 40

The angle ɵ = $\cos^{-1}(\frac{\sqrt{2}}{2})$ = 45°. Since this is a right triangle, it must be isosceles, because we have a 90-degree angle that makes it right, making the third angle 180 – 90 – 45 = 45 degrees also. We know that the ratio of sides for an isosceles right triangle is $1{:}1{:}\sqrt{2}$, so that the hypotenuse *r* must be $30\sqrt{2}$ (the other two sides are both 30). If you do not remember the ratio of sides, use trigonometry as cos(45°) = $\frac{30}{r}$, making *r* = $30\sqrt{2}$. One could also do a quick substitution from the very beginning: since *cos* ɵ = $\frac{\sqrt{2}}{2}$, then $\frac{\sqrt{2}}{2} = \frac{30}{r}$, which will produce *r* = $30\sqrt{2}$.

Answer: A

128. In triangle ABC, AB is extended through D. Find the value of *x*.

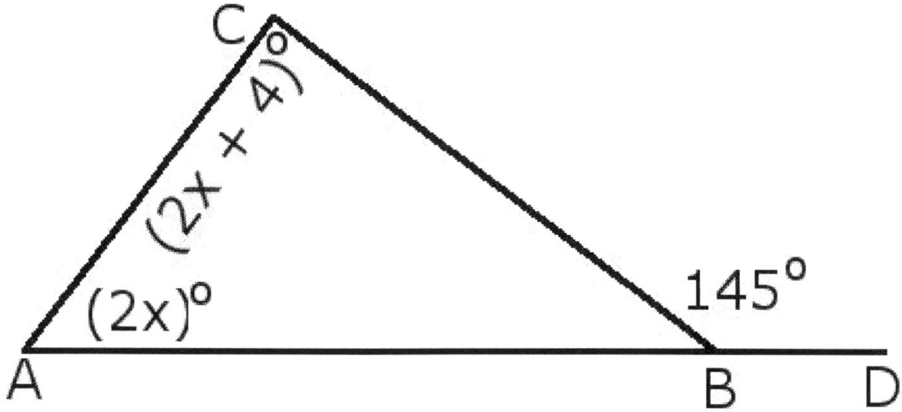

A. 141

B. 145

C. $\frac{145}{4}$

D. $\frac{141}{4}$

Since the exterior angle in a triangle equals the sum of the two non-adjacent interior angles, $145 = 2x + 2x + 4 = 4x + 4$, making $x = \frac{141}{4}$.

Answer: D

129. In the following figure, arc *z*, measuring 40 degrees, is defined by the two equal chords in the circle. A dashed line connects the endpoints defining arc *z*. Find the measure of angle *x*.

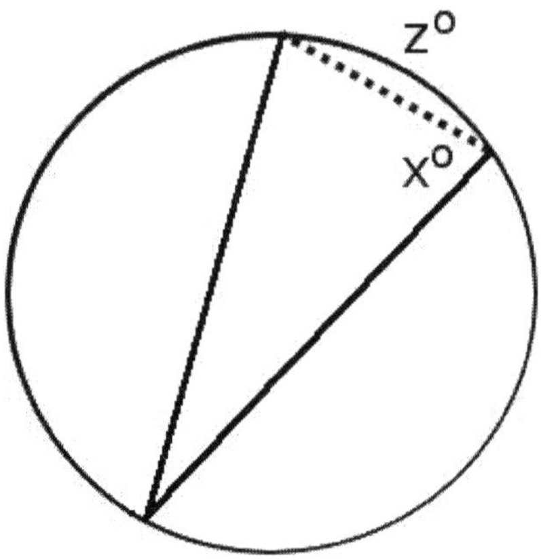

A. 160

B. 180

C. 40

D. 80

Since arc z is spanned by the inscribed angle in a circle, the angle spanning the arc must always be half the degree of the arc itself. Thus, the inscribed angle measures $\frac{40}{2}$ = 20 degrees. Since the two circle chords making this inscribed angle are equal, the triangle created with the dashed line must be isosceles (the dashed line connects the endpoints of the arc z). Thus angle x must measure $\frac{180-20}{2}$ = 80 degrees.

Answer: D

130. A line has slope $\frac{4}{3}$. Express the angle of elevation that makes this slope as a trigonometric function.

A. $\cos^{-1}(\frac{4}{3})$

B. $\sec^{-1}(\frac{3}{5})$

C. $\csc^{-1}(\frac{5}{4})$

D. $\tan^{-1}(\frac{3}{4})$

A slope of $\frac{4}{3}$ means that for every horizontal positive change of 3 there is a vertical positive change of 4. Angle of elevation (or depression) is defined by the steepness of the slope, and it is between the horizontal line and rising (or falling) line (the slope line). We can build a triangle with base 3 (horizontal distance defining the slope), and height 4 (vertical distance defining the slope). The base will be the adjacent side (3) of the angle of elevation, and its opposite side will be the height (4). Hypotenuse (the portion of the slope line) in this triangle must then be $\sqrt{3^2 + 4^2} = 5$. We can avoid the Pythagorean Theorem by simply noting that this is a 3-4-5 triangle. The angle of elevation can be expressed using arccosine, arcsine and arctangent functions. Since the angle with arctangent should be expressed as $\tan^{-1}(\frac{4}{3})$, the last choice is wrong. Choice B is wrong too, because arc secant should define the angle in reference to $\cos^{-1}(\frac{3}{5})$, or $\sec^{-1}(\frac{5}{3})$. The first choice is wrong, because the angle in terms of arccosine should be $\cos^{-1}(\frac{3}{5})$. Choice C makes sense, because arc cosecant defines the angle in reference to $\sin^{-1}(\frac{4}{5})$, or $\csc^{-1}(\frac{5}{4})$.

Answer: C

131. Which of the following represents the *xy*-plane graph below?

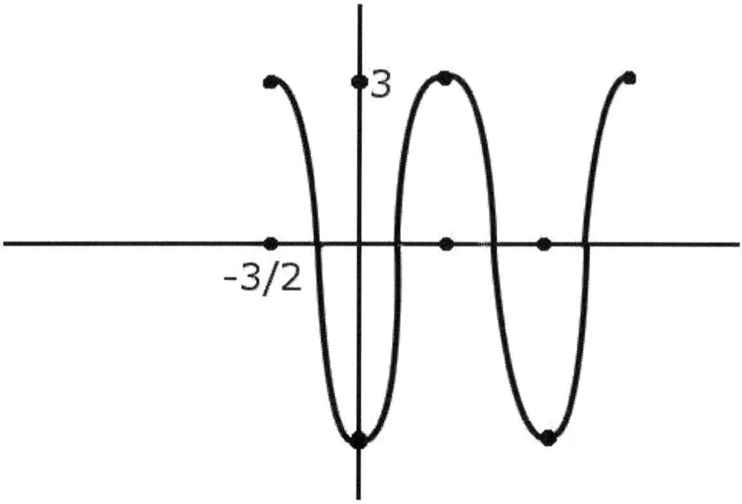

A. $y = -3\cos(\frac{2\pi x}{3})$

B. $y = -3\sin(\frac{2\pi x}{3} + \frac{\pi}{2})$

C. all of the above

D. none of the above

Testing all major *x*-values ($-\frac{3}{2}$, $-\frac{3}{4}$, 0, $\frac{3}{4}$, $\frac{3}{2}$) in choice A corresponds with the graph, and so do *x*-values for choice B. Note that it is always true that $\cos(x) = \sin(x + \frac{\pi}{2})$. It does not matter whether the leading coefficients are positive or negative together in this identity, as long as the coefficient negativity or positivity, and amplitude (3 in this case) are the same.

Answer: C

132. Which *xy*-plane graph below represents equation *y* = -2|*x* – 3| – 2?

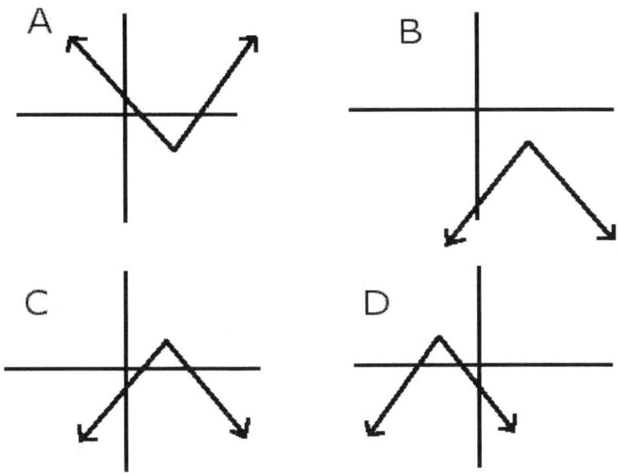

A. graph A

B. graph B

C. graph C

D. graph D

The graph must open down because of the negative leading coefficient, so graph A is ruled out. Since $|x - 3|$ means that there is a 'right' shift of three from the origin, graph D is ruled out. The last indicator -2 after the absolute value bar means that the graph is moved down vertically by 2 after shifting to the right from the origin. This is shown in graph B.

Answer: B

133. What is the area of the rhombus (with two vertices shown) in the figure below?

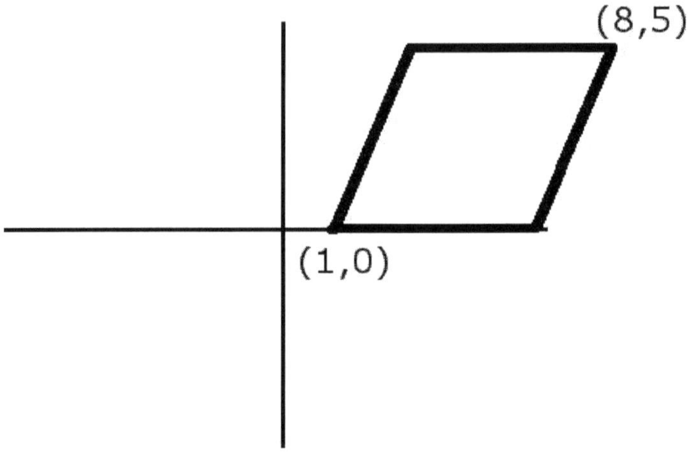

A. 15

B. 14

C. 28

D. cannot be determined from the given information

Since the bottom edge is drawn on the *x*-axis, the upper left corner vertex has a *y*-value of 5. This makes the height of the rhombus 5. Since the rhombus by definition has equal four edges, we can build an equation consisting of the horizontal distance to the upper left corner vertex from the other known vertex (8, 5), and distance from the bottom known vertex (1, 0) to the upper left corner vertex. Using the distance formula we have $\sqrt{(1-x)^2+(0-5)^2} = \sqrt{(8-x)^2+(5-5)^2}$. Getting rid of the square root by squaring, and subsequently simplifying leads to $1 - 2x + x^2 + 25 = 64 - 16x + x^2$, and simplifying further leads to $x = \frac{19}{7}$. This is the *x*-value of the upper left corner vertex of the rhombus. We clearly see the right triangle consisting now of base $\frac{19}{7}$, height 5 and hypotenuse. We need the hypotenuse, because this is one of the equal four edges of the rhombus, thus making it also the base edge of the rhombus that we need in the formula for the area of the rhombus. Using the Pythagorean Theorem we have it as $\sqrt{(\frac{19}{7})^2+5^2}$. The area of the rhombus is then $\sqrt{\frac{361}{49}+25}\,(5) \approx 28$.

Answer: C

134. Calculate the circumference of the circle whose radius matches the outer radius of the torus, if the torus volume is 400 and its inner radius is twice the outer radius.

A. 18

B. 20

C. 12

D. 14

Volume of the torus is calculated by the formula $\pi r^2(2\pi R)$, where r is the outer radius and R inner radius. Since the inner radius is twice the outer radius, this means that $R = 2r$. Substituting this into the volume formula and setting it to 400, we have $400 = 2(2)(\pi)(\pi)(r^2)(r) = 4\pi^2 r^3$. This means that $100 = \pi^2 r^3$, and $r = \left(\frac{100}{\pi^2}\right)^{\frac{1}{3}}$. Since this radius matches the radius of the circle whose circumference we are to find, the circumference

$C = 2\pi\left(\frac{100}{\pi^2}\right)^{\frac{1}{3}} \approx 14.$

Answer: D

135. A fence 10 feet high will be built around a rectangular area that is 200 square feet. If the length of the rectangular area is 25 feet, find the interior area of the vertical space occupied by the fence that surrounds the rectangular area.

A. 660 ft^2

B. 330 ft^2

C. 460 ft^2

D. 230 ft^2

Since the length of the rectangular area is 25, the width must be $\frac{200}{25}$ = 8. Thus, the interior area of the fence material to be built around the rectangular area is 2(8)(10) + 2(25)(10) = 660 square feet.

Answer: A

136. Line segment BC is within the line segment AD. Letters follow each other in alphabetical order from left to right. If BC = 7, AC = 9, and BD = 10, what is the measure of segment AD?

A. 9

B. 10

C. 11

D. 12

According to the problem the line segment (going from left to right) goes in the following order: A, B, C, D. Since AB + BC = AC = 9, and BC = 7, then AB = 9 – 7 = 2. Since BC + CD = BD = 10, and BC = 7, then CD = 10 – 7 = 3. Thus, AD = AB + BC + CD = 2 + 7 + 3 = 12.

Answer: D

137. In an equilateral triangle ABC with side 6 shown in the figure, segment AD is a median. Find the length of segment DE.

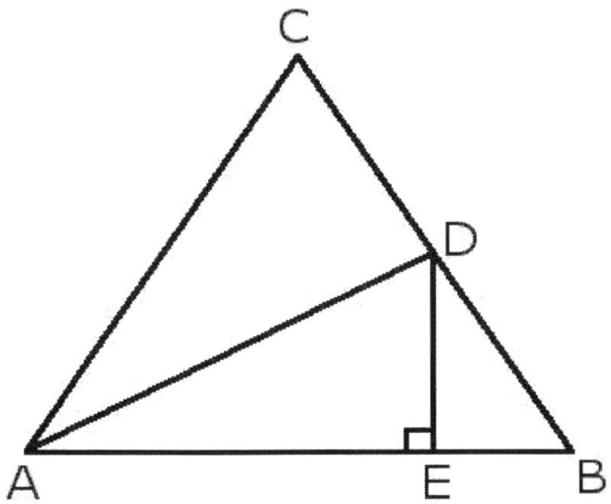

A. $\sqrt{27}$

B. 9

C. $\frac{3\sqrt{3}}{2}$

D. $\frac{9}{2}$

Since this is an equilateral triangle, $m\angle B = 60$ degrees. A median makes DB = $\frac{6}{2}$ = 3. We now have a right triangle BDE with a known side and angle, so we can find DE using trigonometry: $\sin(60^\circ) = \frac{DE}{3}$, so that DE = $\frac{3\sqrt{3}}{2}$.

Answer: C

138. In the following figure, if the area of the isosceles triangle is 104 and base length is $\frac{3}{2}$ the length of the height, find the length of the dashed median (nearest tenth).

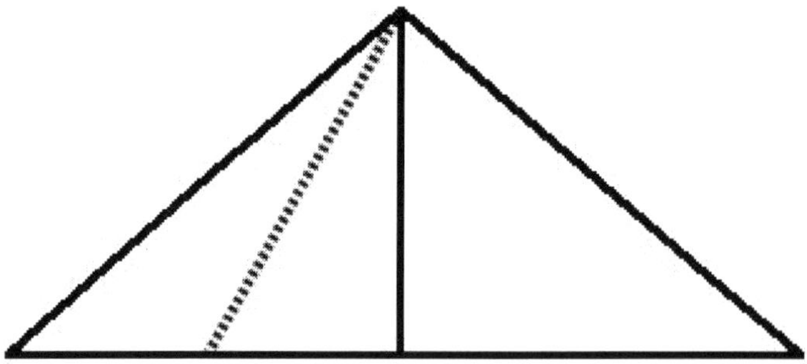

A. 11.8

B. 4.4

C. 12.6

D. 8.8

Since the base $B = (\frac{3}{2})h$, $104 = \frac{\left(\frac{3h}{2}\right)h}{2}$, and $h^2 = \frac{416}{3}$, so that $h = \sqrt{\frac{416}{3}}$. We will find the median as a hypotenuse of the right triangle whose height is h as before, and base $b = \frac{\left(\frac{3}{2}\right)\sqrt{\frac{416}{3}}}{4}$, or simplified as $\frac{3\sqrt{\frac{416}{3}}}{8}$ (we have to divide the original base length by four because besides the height dividing the large base by two we also have the median dividing further by two, so net division by four). Using the Pythagorean Theorem, the median is $\sqrt{\left(\frac{3\sqrt{\frac{416}{3}}}{8}\right)^2 + \sqrt{\left(\frac{416}{3}\right)^2}} \approx 12.6$.

Answer: C

139. If a circle is perfectly inscribed in a square whose diagonal is 10, find the circumference of the circle.

A. $5\sqrt{2}$

B. $\frac{5\sqrt{2}}{2}$

C. $\frac{5\pi\sqrt{2}}{2}$

D. $5\pi\sqrt{2}$

The square side is equal to the diameter of the circle. If we let the square side be *x*, the diagonal will be $\sqrt{x^2 + x^2}$ = $x\sqrt{2}$. Since this equals 10, $x = 5\sqrt{2}$. This is the diameter of the circle. The circumference is then $\pi(5\sqrt{2})$.

Answer: D

140. Find an equation of a line that goes from (-1, 2, 3) to (0, -4, -5) using *t* as a parameter for *x*, *y* and *z*.

A. $x = -t$, $y = -4 + 6t$, $z = 3 - 8t$

B. $x = t - 1$, $y = 2 - 6t$, $z = 3 - 8t$

C. $x = -t$, $y = 2 - 6t$, $z = 3 - 8t$

D. $x = t - 1$, $y = 2 - 6t$, $z = -5 + 8t$

First, we need to find the vector that goes from (-1, 2, 3) to (0, -4, -5). The vector is [0 – (-1), -4 – 2, -5 – 3] = [1, -6, -8]. We then have $x = -1 + t$, $y = 2 - 6t$, $z = 3 - 8t$. The constants before parameter *t* for *x*, *y*, and *z* correspond to the vector. When $t = 0$, the first point on the line is determined, (-1, 2, 3). When $t = 1$, the second point on the line is determined, (0, -4, -5). With increasing *t*, the magnitude (or length) of the line segment increases, while the direction of the line determined by the vector remains the same.

Answer: B

141. Find the volume of the oblique cylinder illustrated in the figure, if its base circle circumference is 29π and slant height 60.

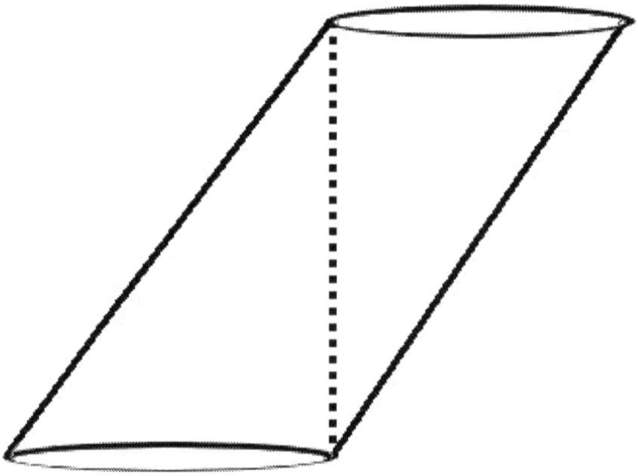

A. 34,695

B. 138,778

C. 39,631

D. 158,525

The base circle circumference of 29π means that $2\pi r = 29\pi$, so that $r = 14.5$. The diameter of the circle base is 29. To find the vertical height, we can construct a right triangle with base the diameter of 29, vertical height x, and hypotenuse that is the slant height of 60. The vertical height is then $x = \sqrt{60^2 - 29^2}$. The volume of the oblique cylinder is the same as the volume of the right cylinder, that is, $\pi(14.5)^2\sqrt{60^2 - 29^2} \approx$ 34,695.

Answer: A

142. If a circle with radius *r* is perfectly inscribed in a square that is circumscribed by a larger circle with radius *R*, the expression that shows the circumference of the large circle in terms of radius *r* is

A. $\sqrt{2}\pi r$

B. $4\sqrt{2}\pi r$

C. $2\sqrt{2}\pi r$

D. $3\sqrt{2}\pi r$

Since each figure is perfectly inscribed in a larger figure circumscribing it, all three figures share the same center point. The small circle radius diameter ($2r$) makes the side of the circumscribing square. The square diagonal is equal to the diameter of the large circle circumscribing it. Since the side of the square is $2r$, its diagonal is $\sqrt{(2r)^2 + (2r)^2} = \sqrt{8r^2} = 2\sqrt{2}r$. As was noted before, this is also the diameter of the large circle. Its circumference is then $\pi(2\sqrt{2}r)$.

Answer: C

143. The circle diameter has endpoints (-1, -6) and (2, 9). The equation of the circle is

A. $(x - \frac{3}{2})^2 + (y - \frac{1}{2})^2 = \sqrt{\frac{117}{2}}$

B. $(x - \frac{1}{2})^2 + (y - \frac{3}{2})^2 = \frac{117}{2}$

C. $(x - \frac{1}{2})^2 + (y - \frac{3}{2})^2 = \sqrt{117}$

D. $(x - \frac{3}{2})^2 + (y - \frac{1}{2})^2 = \frac{117}{2}$

The midpoint of the two given points will be the center of the circle, since the two points are the endpoints of its diameter. The midpoint is $(\frac{-1+2}{2}, \frac{-6+9}{2}) = (\frac{1}{2}, \frac{3}{2})$. The distance between this point and any of the two given points will define the length of the radius of this circle: $\sqrt{\left(-1-\frac{1}{2}\right)^2+\left(-6-\frac{3}{2}\right)^2} = \sqrt{\frac{117}{2}}$. Thus, the equation of the circle is $(x-\frac{1}{2})^2 + (y-\frac{3}{2})^2 = \frac{117}{2}$.

Answer: B

144. If csc θ = $-\frac{13}{5}$, and $\frac{3\pi}{2}$ < θ < 2π, then cot θ is equal to

A. $-\frac{12}{5}$

B. $\frac{5}{12}$

C. $\frac{12}{5}$

D. $-\frac{5}{12}$

Since the angle is in the fourth quadrant in the polar graph of the unit circle, the cotangent function will be negative (cosine is positive and sine is negative), so we can eliminate choices B and C. Since cosecant of the angle equals $-\frac{13}{5}$, the sine of the angle is the reciprocal of that, namely $-\frac{5}{13}$. Since the sine function is opposite divided by the hypotenuse, the opposite side is -5 and hypotenuse is 13. Hypotenuse is always positive, because it is absolute in magnitude and defined by the Pythagorean Theorem, which makes the hypotenuse always positive. The adjacent side will then be $\sqrt{13^2-(-5)^2}$ = 12. Since cotangent of the angle is defined by the adjacent side divided by the opposite side, cot θ = $\frac{12}{-5} = -\frac{12}{5}$.

Answer: A

145. In a standard *x-y* plane, if a point with coordinates (*a*, 0) is reflected through the *y*-axis and then reflected through the origin, which statement below is true?

I. The steps are equivalent to reflecting the original point through the *x*-axis

II. The *y*-coordinate will become negative in the first translation

III. The *x*-coordinate is negated in the final translation

IV. The *y*-coordinate becomes *a* in the final translation

A. I and IV

B. I, II and III

C. I only

D. I and III

The first statement is true. The first step transforms the point (*a*, 0) into (-*a*, 0), and the second step transforms the point (-*a*, 0) into (*a*, 0). If one reflects the original point (*a*, 0) through the *x*-axis, the only value to change negatively would be the *y*-value, which is zero. Thus, the reflection would produce the same point (*a*, 0), unchanged. Statement II is false, because only the *x*-value becomes negative if a point is reflected through the *y*-axis. Statement III is true, because the *x*-coordinate is negated when reflected through the origin, so being negative in the second quadrant it becomes positive in the fourth quadrant. Statement IV is false: the *y*-coordinate would only become *a* if a point was in the function that was inverted, which is not the case here.

Answer: D

146. If a scalene triangle has vertices (1, 0), (5, 0) and (4, 3), find its smallest interior angle.

A. 37°

B. 18°

C. 117°

D. 20°

It helps to draw this triangle in a standard Cartesian *x-y* plane. The smallest angle in this triangle is at the point (1, 0), and this should come from the visual understanding of the graph drawn, and also from the fact that the opposite side length of any given angle is proportional to the angle measure. In other words, the shortest side is drawn from point (5, 0) to point (4, 3), so that the angle opposite of this side is the smallest in measure. A right triangle may be constructed to find this angle, with opposite side being 3 in length, and adjacent side being 4 in length. We know this from the point (4, 3) while noting its coordinates. The angle measure is then $\tan^{-1}(\frac{3}{4}) \approx 37$ degrees.

Answer: A

147. A triangle is defined to be bounded by three equations: $x = 0$, $y = x$ and $y = -x + 8$. Choose the best answer that describes this triangle.

A. Scalene and right

B. Isosceles and right

C. Isosceles and non-right

D. Scalene and non-right

To understand what the graph might look like, we first find the *x*-value of the intersection of the lines $y = x$ and $y = -x + 8$, by setting the *y*-values equal: $x = -x + 8$, giving $x = 4$. Since the base of the triangle is defined by the *y*-axis (where $x = 0$), this means both equations start from the point where they are 8 units apart (one equation has $y = 8$, the other $y = 0$ at $x = 0$). This is the base length of the triangle. The slopes of the two lines are the negative reciprocals of each other, so the lines are mutually perpendicular. This rules out choices C and D. Since the lines intersect at the midpoint of the base ($y = 4$ at $x = 4$), with each line covering exactly half their starting vertical distance of 8 (base length), this makes the triangle isosceles and rules out choice A.

Answer: B

148. An equation of a line that is perpendicular to equation $y = (\frac{2}{5})x - 3$ and passes through the point (3, -12) is

A. $y = \frac{5}{2}x + 27$

B. $y = -\frac{5}{2}x + 27$

C. $y = -\frac{5}{2}x + \frac{39}{2}$

D. $y = -\frac{5}{2}x - \frac{9}{2}$

The slope of the perpendicular line is the negative reciprocal to the slope of the given line, that is, $-\frac{5}{2}$. Using the given point through which the perpendicular line to the given line passes, we have $y = -\frac{5}{2}(x - 3) - 12 = -\frac{5}{2}x + \frac{15}{2} - 12 = -\frac{5}{2}x - \frac{9}{2}$.

Answer: D

149. Choose the best equation of a line that will make a scalene right triangle with the two lines whose equations are $y = \frac{3}{4}x + 6$ and $x = 0$.

A. $y = -\frac{4}{3}x - 6$

B. $y = \frac{4}{3}x$

C. $y = 0$

D. $y = \frac{4}{3}x + 6$

We can immediately eliminate choices B and D. Choice B will not give a right triangle with the first given equation - the slope of $\frac{4}{3}$ is not parallel to either the x-axis or the y-axis, and the line makes an acute angle at the point of intersection with the first given line; choice D makes the two equations start at the same point $y = 6$, and there is not even a boundary seen as *x* moves to the right, that will make a triangle with the two equations let alone a right triangle. Choice A qualifies for a right triangle, but it produces an isosceles triangle, because the *y*-intercepts of the two equations are exactly 12 units apart, with the origin being the midpoint of the triangle base in this triangle. Thus, only Choice C makes a scalene right triangle with all distinct sides, and with two sides consisting of lines $y = 0$ and $x = 0$ that are mutually perpendicular.

Answer: C

150. If points A, B and C are located on the same straight line, with AB = 12 and AC = 5, what is the positive difference of all possible lengths of BC?

A. 17

B. 7

C. 10

D. 5

According to the values given, two possible lines may exist, and the order of the letters for each line from left to right is ACB and CAB. In the first line, BC = AB – AC = 12 – 5 = 7. For the second line, BC = AB + AC = 12 + 5 = 17. Thus, the positive difference between the two lengths of BC we found is 17 – 7 = 10.

Answer: C

151. Find the *x*-value(s) for the point(s) at which the lines of the equations

$y = -(x - 4)^2 - 2$ and $y = (x - 3)^2 - 5$ intersect.

A. $\frac{7}{2} - \frac{\sqrt{5}}{2}$

B. $\frac{7}{2} + \frac{\sqrt{5}}{2}$

C. $\frac{7}{2} \pm \frac{\sqrt{5}}{2}$

D. $\frac{7}{4} \pm \frac{\sqrt{5}}{4}$

When any two lines intersect, their *y*-values are equal (and *x*-values as well). We set the two equations equal (their *y*-values): $-(x-4)^2 - 2 = (x-3)^2 - 5$. Simplifying and combining the like terms, we have a quadratic equation $2x^2 - 14x + 22 = 0$. Solving for *x* using the quadratic formula (where $a = 2$, $b = -14$, and $c = 22$), we get $x = \frac{7}{2} \pm \frac{\sqrt{5}}{2}$.

Answer: C

152. A page whose width is $\frac{1}{4}$ its length has side margins whose length is three times the length of the vertical margins. If the combined area of the space occupied by the side and vertical margins is 300, what is the value of the length of the page when the side margin is 6?

A. 51

B. 53

C. 57

D. 50

It always helps to draw a picture when it is not accompanying the problem. If we let the length of the page be *y*, the width $(\frac{1}{4})y$, the vertical margin *x*, and the side margin 3*x*, we will have an interesting figure to work with. Take a look at the figure below.

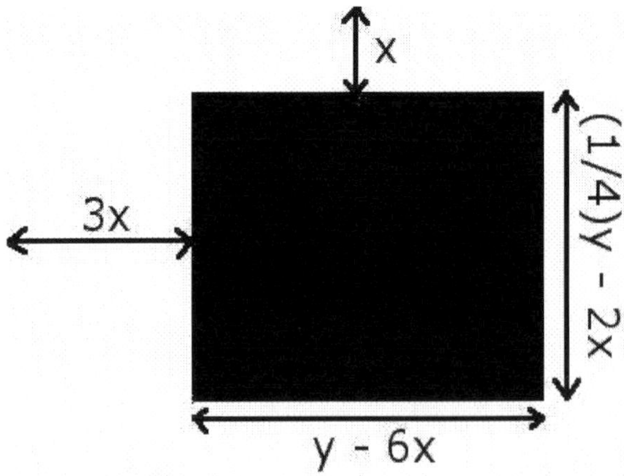

Since the space occupied by the margins is 300, we have $300 = 2(\frac{3xy}{4}) + 2x(y - 6x)$, which simplifies to $300 = \frac{7xy}{2} - 12x^2$. We got this by adding the areas defined by the width and side margin (two such areas) and the remaining area of the space occupied by the vertical margins and length part (two such spaces). Since we are asked to find value of the length of the page given the side margin of 6, we need to rewrite the equation we found so that y is in terms of x. After doing a bit of algebra when isolating the y on one side, we have $y = \frac{600}{7x} + \frac{24x}{7}$. Since the side margin is 6, $3x = 6$, which makes $x = 2$. Using this value of x we get $y = \frac{600}{14} + \frac{48}{7} \approx 50$.

Answer: D

153. A ball is thrown vertically upward from the ground level with an initial velocity of $x = 60$ feet per second. Its vertical position at time t is given by $s_t = -16t^2 + xt + s_0$. In how many seconds does it hit the ground?

A. 16

B. $\frac{15}{4}$

C. 4

D. 5

In this problem $x = 60$ is the initial velocity (positive, because the ball is thrown upward), s_o is the starting vertical point of the ball at time zero, and s_t is the final vertical point of the ball at time *t*. Since the starting vertical point at time zero is 0, $s_0 = 0$, and we also set the final position of the ball zero, so that $s_t = 0$ (we want to know the time when the ball hits the ground). Thus we have $0 = -16t^2 + 60t + 0$, and factoring the expression we get $0 = -16t(t - \frac{60}{16})$, so that we have two times when the ball is at the zero vertical position: $t = 0$ (this makes sense because the ball is fired at the ground level) and $t = \frac{15}{4}$. Thus the ball hits the ground in $\frac{15}{4}$ seconds.

Answer: B

154. A rectangular field with an area of 200 square meters should have a maximum perimeter of 100. What is the possible range of the values for the length of the field?

A) $25 - 5\sqrt{17} < L \leq 25 + 5\sqrt{17}$

B) $25 < L \leq 25 + 5\sqrt{17}$

C) $10\sqrt{2} < L \leq 25 + 5\sqrt{17}$

D) $10\sqrt{2} < L \leq 25$

If we let *x* be the length, and *y* be the width of the rectangular field, then $xy = 200$, and we must have $2x + 2y \leq 100$, or $x + y \leq 50$. Since $y = \frac{200}{x}$, then $x + \frac{200}{x} \leq 50$, which makes $x^2 + 200 - 50x \leq 0$. In standard quadratic form this is $x^2 - 50x + 200 \leq 0$. Using quadratic formula we arrive at the two roots $25 \pm 5\sqrt{17}$, which means that $[x - (25 + 5\sqrt{17})][x - (25 - 5\sqrt{17})] \leq 0$, so that $25 - 5\sqrt{17} \leq x \leq 25 + 5\sqrt{17}$, or theoretically about

$4.38 \leq x \leq 45.62$. However, you have to note that the length must always exceed the width, and since length multiplied by width is 200, then length $x > \sqrt{200} = 10\sqrt{2}$. Thus the range for the length must be $10\sqrt{2} < L \leq 25 + 5\sqrt{17}$.

Answer: C

155. Find an equation of the tangent line to the circle $(x - 3)^2 + (y + 2)^2 = 100$ at the point (-3, 6).

A. $y = \frac{3}{4}x + \frac{15}{24}$

B. $3x - 4y = -33$

C. $y = -\frac{4}{3}x - \frac{15}{24}$

D. $18x - 24y = -15$

The equation of the tangent line will be perpendicular to the radius of the circle at the point (-3, 6). Since the center of the circle is at (3, -2), the slope of the radius line is $\frac{-2-6}{3-(-3)} = \frac{-8}{6} = -\frac{4}{3}$. Thus, the tangent line will have a slope of $\frac{3}{4}$ (the negative reciprocal of the radius line). The equation of the tangent line is then $y = \frac{3}{4}(x + 3) + 6$, or $y = \frac{3x}{4} + \frac{33}{4}$. Rewriting the equation we have $4y = 3x + 33$, or $3x - 4y = -33$.

Answer: B

156. In the following figure, two right triangles ABC and DBE are juxtaposed together such that CB is perpendicular to DE. If $_m$<DEB = 32°, find the measure of angle *C*.

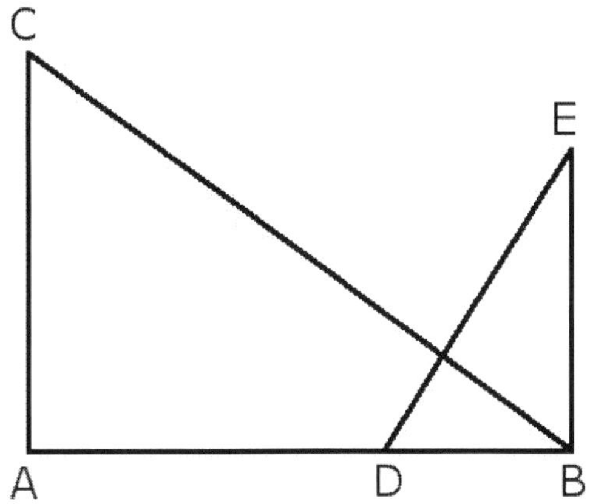

A. 58°

B. 56°

C. 54°

D. 52°

Since $_m$<DEB = 32 degrees, and <DBE is a right angle, $_m$<EDB = 180 – 32 – 90 = 58°. This makes $_m$<CBA = 180 – 90 – 58 = 32° (CB and DE make a right angle because they are perpendicular to each other. Since triangle ABC is a right triangle, $_m$<C = 180 – 32 – 90 = 58 degrees. Note that you could avoid all this work after you found $_m$<EDB = 58° by simply noticing that since CA is perpendicular to AB and CB is perpendicular to DE then $_m$<EDB = $_m$<C.

Answer: A

157. If $x = \sec^{-1}(\frac{5}{3})$, and $0 \le x \le \frac{\pi}{2}$, what is the value of $[\sin(x) + \cos(x)]^2$

A. $\frac{37}{25}$

B. $\frac{12}{25}$

C. $\frac{24}{25}$

D. $\frac{49}{25}$

From the definition of secant given, we see that $x = \cos^{-1}(\frac{3}{5})$, so that $\cos(x) = \frac{3}{5}$, since we know that $\cos(x) = \frac{1}{\sec(x)}$. Since 5 is a hypotenuse, and we are in quadrant I in the unit circle, $\sin(x) = \frac{4}{5}$. We get the denominator 4 either from the Pythagorean Theorem or by simply noting the 3-4-5 triangle sides. Thus, $[\sin(x) + \cos(x)]^2 = (\frac{4}{5} + \frac{3}{5})^2 = (\frac{7}{5})^2 = \frac{49}{25}$.

Answer: D

158. In a triangle, a hypotenuse is three times the length of one of its legs. This triangle is therefore

A. equilateral

B. isosceles

C. oblique

D. scalene

Since a hypotenuse is mentioned, choice C is irrelevant. Hypotenuse also means that choice A is irrelevant (even without noting that the hypotenuse is three times the length of one of its legs). If a hypotenuse is $3x$, and one leg is x, the third leg is $\sqrt{9x^2 - x^2} = \sqrt{8x^2} = 2\sqrt{2}x$. Thus, all sides of this right triangle are distinct. Therefore it is scalene.

Answer: D

159. If AB and CD are diameters of the circle shown, and angle *z* is 128 degrees, what is the measure of angle *y*?

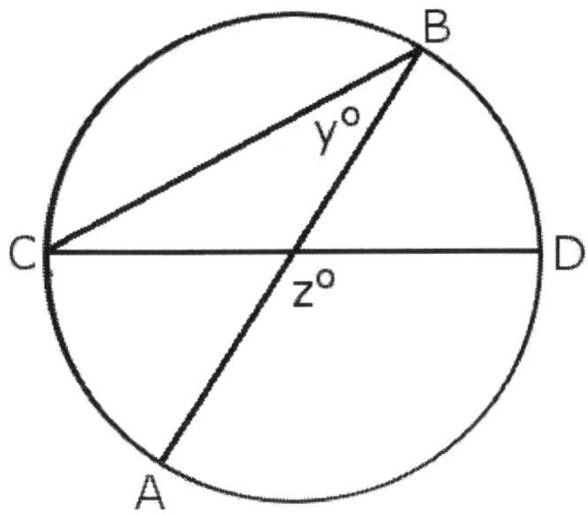

A. 64°

B. 52°

C. 26°

D. 32°

The vertical angle of *z* (also a central angle) also measures 128 degrees, and this is a vertex angle of an isosceles triangle with one of its equal angles being *y*. The triangle is isosceles because two of its sides are radiuses of the circle. Thus, angle *y* measure is $\frac{180-128}{2}$ = 26 degrees. We could have found it by first finding the supplementary angle with *z*, which is 180 – 128 = 52 degrees in measure. This is a central angle. Since angle *y* is an inscribed angle that also spans the same arc as the central angle of 52 degrees, it must be half the measure of the arc, that is, $\frac{56}{2}$ = 26 degrees.

Answer: C

160. In the following figure, if the area of the right triangle ABC is 3, and BC = 2AB, find the length of the median AD. (figure not drawn to scale).

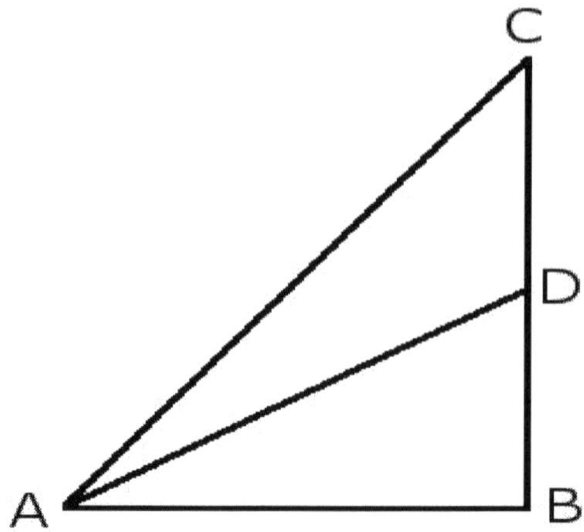

A. $9\sqrt{2}$

B. $2\sqrt{6}$

C. $\sqrt{6}$

D. $\frac{\sqrt{6}}{2}$

The area of the triangle ABC is 3 = $\frac{AB(2AB)}{2}$, so that AB = $\sqrt{3}$. Since BC = 2AB, and the median AD makes CD = BD, then BD = AB = $\sqrt{3}$. Thus AD = $\sqrt{(\sqrt{3})^2 + (\sqrt{3})^2}$ = $\sqrt{6}$.

Answer: C

161. A full aquarium with volume 90 mL is being emptied into another aquarium with length 6 cm and width 7 cm. What is the height of this aquarium?

A. the other aquarium will not take this much water

B. greater than 3 centimeters

C. between 2 and 3 centimeters

D. less than 2 centimeters

We know that 1 cm^3 = 1 mL. If we let the height of the new aquarium be x centimeters, then $90 = 6(7)(x) = 42x$, so that $x = \frac{90}{42}$ centimeters. This is greater than 2 centimeters, but less than 3 centimeters.

Answer: C

162. The volume of the cube is 300. The longest line segment connecting the two vertices of the cube is about

A. 12

B. 11

C. 10

D. 9

If we let each side of the cube be x, then $300 = x^3$, so that $x \approx 6.69$. The longest line segment connecting the two vertices of the cube is the longest diagonal of the cube, which covers three dimensions. Thus, it must be $\sqrt{x^2 + x^2 + x^2} = \sqrt{3x^2} = x\sqrt{3} = 6.69\sqrt{3} \approx 12$.

Answer: A

163. A current square-shaped pool has a walkway with uniform width surrounding it. A diagonal fence is placed at the two corners of the pool, extending through the walkway (see figure). A new renovation project suggests building square-shaped baths around the pool, and the two corners will serve as staircase exits from the pool. If the area of the pool is 64 and the area of the pool plus the walkway is 100, how many complete 1 x 1 baths can be built on the surrounding walkway?

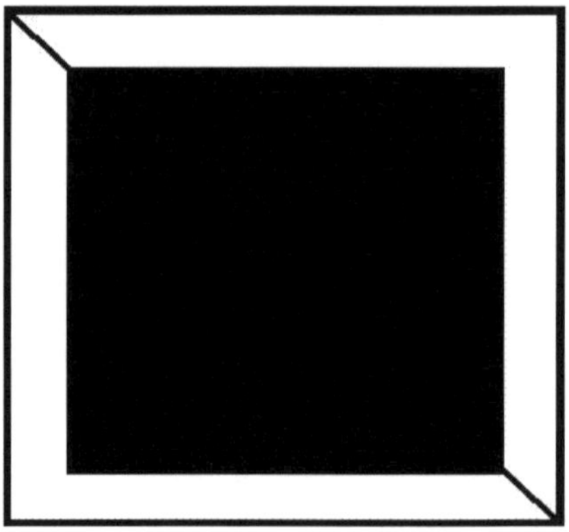

A. 32

B. 33

C. 34

D. 35

Since the area of the entire region including the walkway is 100, each walkway side is 10. Since each pool side is 8, there can be one 1 by 1 bath at each of the two corners where there is no staircase (where there is no diagonal line drawn). Since each pool side length can fit eight 1 x 1 baths, the total number of baths possible is 4(8) + 1 + 1 = 34.

Answer: C

164. If a rhombus is made up of two equilateral triangles with side 6, what is its area?

A. 36

B. 18

C. $18\sqrt{3}$

D. cannot be determined from the given information

We can easily determine the height of the rhombus, which is also the height of any of the two equilateral triangles: if we break the equilateral triangle in half, we will break the triangle base in half as well (the base will be $\frac{6}{2}$ = 3).

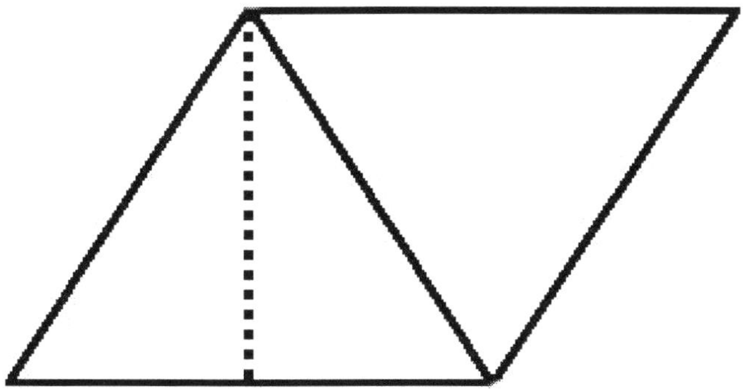

The right triangle is formed with hypotenuse 6, base 3, and a height that is $\sqrt{6^2 - 3^2}$ = $\sqrt{27}$ = $3\sqrt{3}$. Since the base of the rhombus is 6, its area is $6(3\sqrt{3})$ = $18\sqrt{3}$.

Answer: C

165. Equation of the line $-\frac{3}{4}x - \frac{2}{3}y + 6 = -23$ has an *x*-intercept at

A. $x = -\frac{116}{3}$

B. $x = \frac{87}{2}$

C. $x = \frac{116}{3}$

D. $x = \frac{68}{3}$

An x-intercept occurs when $y = 0$. Thus, $-\frac{3x}{4} - 0 + 6 + 23 = 0$, and $x = \frac{(-29)(4)}{-3} = \frac{116}{3}$.

Answer: C

166. If a point (x, y) is located on the unit circle (with radius 1) and defined by an angle of 135 degrees, then $x^2 - y^2$ is

A. 1

B. 4

C. -1

D. 0

Since the point (x, y) is defined by a terminal side making a 135-degree angle, this means that the point is in quadrant II of the unit circle. Since a 135-degree angle is equivalent to $\frac{3\pi}{4}$ in radians, the point (x, y) is defined as $(\cos(\frac{3\pi}{4}), \sin(\frac{3\pi}{4}))$, or $(-\frac{\sqrt{2}}{2}, \frac{\sqrt{2}}{2})$. Thus, $x^2 - y^2 = \frac{1}{2} - \frac{1}{2} = 0$.

Answer: D

167. In the following figure, an isosceles triangle shares a leg with a side of a square. Find the measure of angle x.

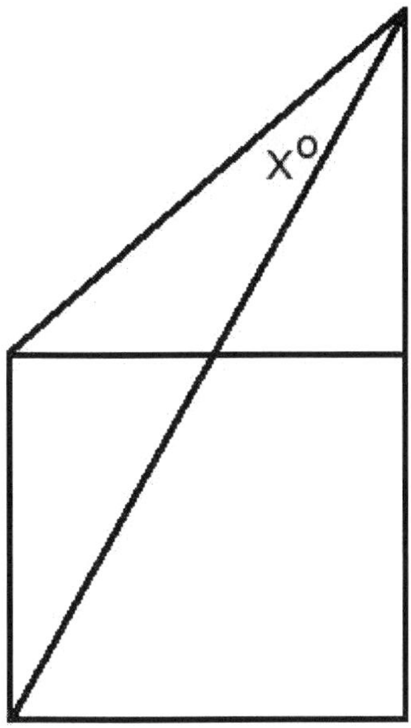

A. 45°

B. 27°

C. 18°

D. cannot be determined from the given information

The isosceles triangle has a 45-degree angle that consists of angle x and its adjacent angle. Let us name the adjacent angle y. If we let the square side be a, this square side a is also one of the identical sides of the isosceles triangle. A large right triangle whose base is a square side a and height $2a$ has angle y whose opposite side is a and adjacent side $2a$. Thus, angle y is $\tan^{-1}(\frac{a}{2a}) = \tan^{-1}(\frac{1}{2}) \approx 26.57$ degrees. This makes angle $x = 45 - 26.57 \approx 18$ degrees.

Answer: C

168. A hyperbola equation is

A. $(x - 2)^2 + (y + 4)^2 - \sqrt{101} = 0$

B. $(x + 4)^2 - y^2 = 0$

C. $36x^2 - 25y^2 + 216x + 100y - 676 = 0$

D. none of the above

The key here is to first see one positive coefficient and one negative coefficient in front of either x or y. Choice A is thus automatically ruled out (it is a circle). Choice B has the hyperbola coefficient variety characteristics, but if one rewrites the equation it is easy to see that the graph will be $y = \pm(x + 4)$ which is not a hyperbola (it is a graph of two perpendicular lines). Choice C could work, but we still have to check it by rewriting the equation so that the right side equals 1. First we have $36x^2 + 216x - 25y^2 + 100y = 676$. By completing the square we have $36(x^2 + 6x + 9) - 324 - 25(y^2 - 4y + 4) + 100 = 676$, which simplifies as $36(x + 3)^2 - 25(y - 2)^2 = 900$. Thus, we have $\frac{(x+3)^2}{25} - \frac{(y-2)^2}{36} = 1$, and this is definitely an equation of a hyperbola.

Answer: C

169. A random two-dimensional shot captures a right circular cone as an isosceles triangle. If its vertex angle measures 43 degrees, and slant height measures 10, find the volume of the cone.

A. 332

B. 131

C. 201

D. 71

If we divide the isosceles triangle in half, we get two equal right triangles with the same height and the 43-degree vertex angle is split in half, so it becomes 21.5 degrees. Working with one of these two right triangles, the height h is found using the slant height and the angle of 21.5 degrees: $\cos(21.5^\circ) = \frac{h}{10}$, making the height $h \approx 9.304$. The base of this triangle is the radius of the cone, and is $\sqrt{10^2 - (9.304)^2} \approx 3.665$. Thus, the volume of the cone is $\frac{\pi((3.665)^2)(9.304)}{3} \approx 131$.

Answer: B

170. A triangle has dimensions 8*x*, 15*x*, and 17*x*. The difference of the two interior angles in this triangle can be about

A. 90°

B. 34°

C. 25°

D. 65°

Since this is a 8-15-17 right triangle, we know that one of the angles (the largest angle) measures 90 degrees. Using the *Law of Sines*, we have $\frac{\sin(90^o)}{17} = \frac{\sin(x)}{15}$, which gives $x = \sin^{-1}(\frac{15}{17}) \approx 62$ degrees. The third angle then measures about $180 - 62 - 90 \approx 28$ degrees. Choosing among the three possible interior angle differences, we have $62 - 28 = 34$ degrees that is listed in the answer choices provided.

Answer: B

171. A triangle drawn from the origin has the other two vertices at (2, -3, -4) and (0, 0, -6). The greatest interior angle in this triangle measures about

A. 61°

B. 90°

C. 77°

D. 42°

One side of the triangle is 6. We know this from the coordinates (0, 0, -6). Using (2, -3, -4), the second side of the triangle is $\sqrt{(2-0)^2+(-3-0)^2+(-4+6)^2} = \sqrt{17}$. Using the coordinates of the origin and (2, -3, -4), we have the third side as $\sqrt{(2-0)^2+(-3-0)^2+(-4-0)^2} = \sqrt{29}$. The greatest interior angle will have the longest side of the triangle as its opposite side. This is side with length 6. Using the *Law of Cosines*, we have $6^2 = \left(\sqrt{29}\right)^2 + \left(\sqrt{17}\right)^2 - 2(\sqrt{17})(\sqrt{29})\cos(x)$, so that $x = \cos^{-1}(0.225188675) = 77$ degrees.

Answer: C

172. A random two-dimensional shot shows a square perfectly inscribed in a circle so that the square diagonal equals the diameter of the circle. If this is actually a cube with side 3 (shown in R^3), inscribed in a sphere, find the volume portion of the sphere not including the cube.

A. $(\frac{27\sqrt{2}}{2})(\frac{4\pi}{3} - 1)$

B. $(27\sqrt{2})(\frac{4\pi}{3} - 1)$

C. $\frac{9\pi\sqrt{2}}{2} - 27$

D. $9\pi\sqrt{2} - 27$

Working with a square, we have its diagonal as $\sqrt{3^2+3^2} = 3\sqrt{2}$. This is the diameter of the circle (and the sphere). The radius of the circle and sphere is then $\frac{3\sqrt{2}}{2}$. Cube volume is $3^3 = 27$. Sphere volume is $4\pi\frac{\left(\frac{3\sqrt{2}}{2}\right)^3}{3} = 9\pi\sqrt{2}$. Thus, the volume of the sphere portion not taken by the enclosed cube is $9\pi\sqrt{2} - 27$.

Answer: D

173. The following figure shows an equilateral triangle with side 5 that is being divided into three equal triangles sharing a vertex point. What is the area of the shaded region?

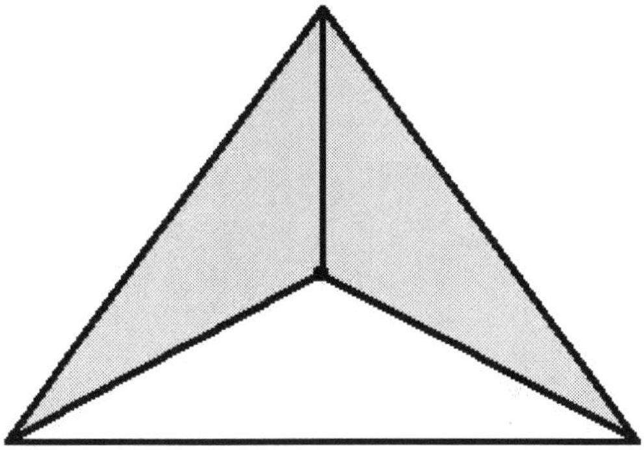

A. $25\sqrt{3}$

B. $\frac{25\sqrt{3}}{2}$

C. $\frac{25\sqrt{3}}{4}$

D. $\frac{25\sqrt{3}}{6}$

Since the three triangles are equal, one simply needs to find $\frac{2}{3}$ of the area of the equilateral triangle, which matches the combined area of the two equal triangles. Area of the equilateral triangle is $\frac{5^2\sqrt{3}}{4} = \frac{25\sqrt{3}}{4}$. Thus, the area of the two inscribed triangles is $(\frac{25\sqrt{3}}{4})(\frac{2}{3}) = \frac{25\sqrt{3}}{6}$.

Answer: D

174. In the following figure, a rhombus is inscribed in a rectangle. Rhombus vertices divide the length and width of the rectangle in half. Suppose this figure is drawn in a standard *x-y* coordinate plane, and the coordinates of the two opposite rectangle vertices are (-11, -17) and (5, -6). What is the area of the rhombus?

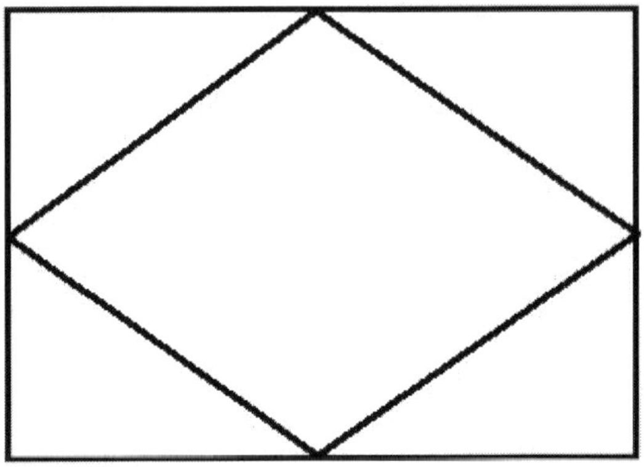

A. 66

B. 88

C. 176

D. 132

Since the rhombus vertices divide the length and width of the rectangle in half, diagonals of the rhombus are parallel to the length and width of the rectangle. If we take one coordinate of the rectangle vertex (-11, -17), the vertex on the same line (length of the rectangle) with this point must be on (5, -17). Thus, the length of the rectangle is |-11| + 5 = 16, and width is |-17 –(-6)| = |-17 + 6| = |-11| = 11. Since the area of the inscribed rhombus is half the area of the rectangle (rhombus diagonals are parallel to the length and width of the rectangle), the rhombus area is $\frac{11(16)}{2}$ = 88.

Answer: B

175. How many complete isosceles triangles, with each having two sides measuring 1.5, can fit inside the rectangle with dimensions 3 and 7?

A. 16

B. 17

C. 18

D. 19

Since the question requires finding the number of complete triangles, this means that there may or may not be extra space after the complete triangles exhaust the rectangle space. We can simply count the number of squares with side measuring 1.5, each of which has two identical triangles with two sides measuring 1.5. The rectangle length of 7 can fit 4 such squares, and rectangle width of 3 can fit exactly 2 such squares. Thus, there are 2(4) = 8 squares in total that fit into the rectangle, which means that 8(2) = 16 isosceles triangles may fit into the rectangle. This takes 18 units out of 21 units of the total rectangle area. Even though each triangle area is $\frac{(1.5)^2}{2}$ = 1.125, and at least two more triangle areas would theoretically fill the remaining space, the space itself does not allow any more than 16 triangles to fit, because we have a 1 by 3 rectangle as the extra space.

Answer: A

176. In the following figure showing an equilateral triangle, AB and CD are medians. If the area of the equilateral triangle is 25, what is the area of ΔACD (not shown)?

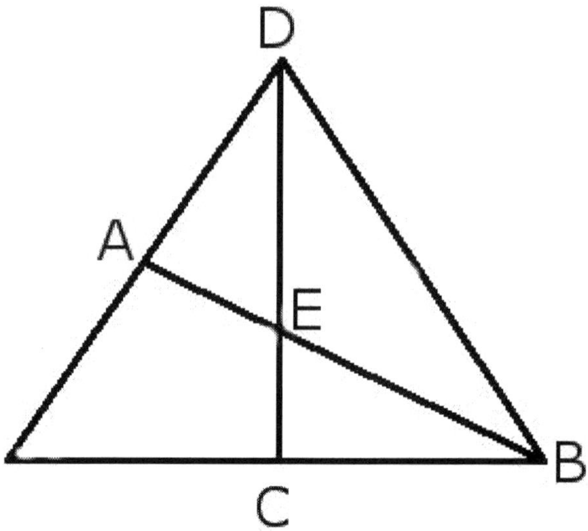

A. 7.3

B. 6.3

C. 12.6

D. 14.4

The area of the triangle is 25 = $\frac{s^2\sqrt{3}}{4}$, so that its side s = $10(3^{-\frac{1}{4}})$ ≈ 7.598, so that AD = $\frac{7.598}{2}$ = 3.799. We need to find the measure of median CD, because CD is a base for the triangle whose area we need to find. Also, BC = AD, because AB and CD are equal medians (equilateral triangle medians). Using the Pythagorean Theorem on BC and DB, CD = $\sqrt{(7.598)^2 - (3.799)^2}$ ≈ 6.58. The perpendicular height *h* from CD to A is found by triangle similarity and direct proportion associated with it: $\frac{AD}{s} = \frac{h}{3.799}$, so that becomes $\frac{1}{2} = \frac{h}{3.799}$, making *h* ≈ 1.9. Thus, the area of ΔACD is $\frac{(1.9)(6.58)}{2}$ ≈ 6.3.

Answer: B

177. Nine identical circles are placed in a row. A line is drawn through the center of these circles so that it covers the combined distance of their diameters. What is the minimum number of these lines needed to satisfy the combined distance of revolutions of these nine circles if a distance of three revolutions of each circle must be satisfied?

A. 8

B. 9

C. 10

D. 11

If we let the diameter of each of the nine circles be 1, the radius of each of the nine circles is $\frac{1}{2}$. The distance covered by the line through the center of these nine circles is then 9. Three revolutions of each circle is $\frac{3(2\pi)}{2} = 3\pi$. The combined distance of three revolutions for all nine circles is $3\pi(9) = 27\pi$. Thus, the minimum number of lines with length 9 needed to satisfy the combined distance of all nine circles with three revolutions is $\frac{27\pi}{9} = 3\pi \leq 10$.

Answer: C

178. Find the asymptotes of a hyperbola whose equation is $5x^2 - 6y^2 - 30 = 0$.

A. $\frac{6x}{5}, -\frac{6x}{5}$

B. $\frac{\sqrt{6}x}{\sqrt{5}}, -\frac{\sqrt{6}x}{\sqrt{5}}$

C. $\frac{\sqrt{5}x}{\sqrt{6}}, -\frac{\sqrt{5}x}{\sqrt{6}}$

D. $\frac{5x}{6}, -\frac{5x}{6}$

Putting the equation in standard hyperbola form, we have $\frac{x^2}{6} - \frac{y^2}{5} = 1$. Since the y^2 coefficient is negative and the center is at (0, 0), the hyperbola's foci are on the horizontal line, precisely on the *x*-axis. This means that the asymptotes for this hyperbola will be the lines $\pm (\frac{b}{a})x$. Since $a = \sqrt{6}$ and $b = \sqrt{5}$, the asymptotic lines are $\pm \sqrt{\frac{5}{6}}x$.

Answer: C

179. Find the area of a region bounded by the *x*-axis and lines *y* = *x* + 5, *x* = 3.

A. 8

B. 64

C. 16

D. 32

The *y*-value of *y* = *x* + 5 is zero at *x* = -5, and eight at *x* = 3. The other two lines are vertical and horizontal (*x*-axis and *x* = 3). The hypotenuse is determined by the line *y* = *x* + 5, so that the area of the triangle created by the three lines is $\frac{(|-5| + 3)(8)}{2} = 32$.

Answer: D

180. What is the approximate slope of the line that is perpendicular to the collinear line with the longest oblique triangle side shown?

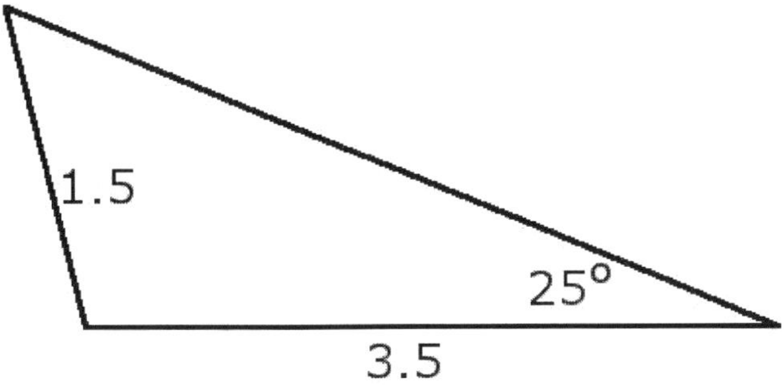

A. 1.97

B. 3.55

C. 2.99

D. 2.37

First we find the angle opposite of which is the side measuring 3.5: using the *Law of Sines*, we have $\frac{\sin(25^o)}{1.5} = \frac{\sin(x)}{3.5}$, and $x \approx 80.44$ degrees. The longest will make a hypotenuse is the triangle is a right triangle. Thus, the complementary angle with the one we just found is 90 – 80.44 = 9.56 degrees. The horizontal distance needed to be added to 3.5 in order to create a right triangle is *d*: $\sin(9.56^o) = \frac{d}{1.5}$, giving $d \approx 0.249$. This makes the base of the right triangle 0.249 + 3.5 = 3.749. The vertical height of the right triangle is $\sqrt{(1.5)^2 - (0.249)^2} \approx 1.479$. This makes the slope of the line collinear with the hypotenuse of the right triangle $\frac{-1.479}{3.5} \approx -0.4226$. The slope of the perpendicular line to the hypotenuse is then $\frac{-1}{-0.4226} \approx 2.37$.

Answer: D

181. Find the area of the shaded region below if the radius of the circle shown is 5.

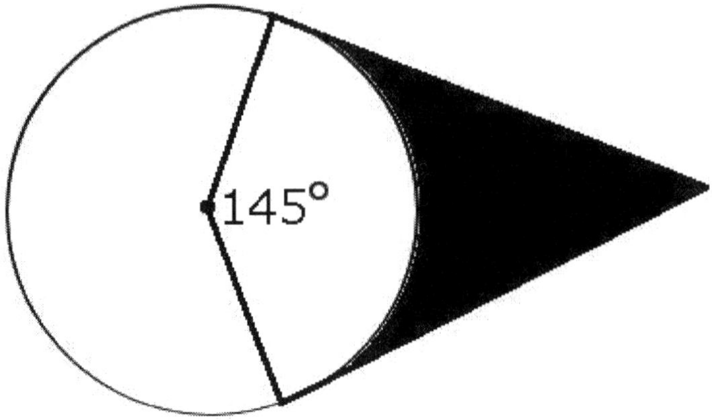

A. 8

B. 24

C. 48

D. cannot be determined

Area of the shaded region is determined by subtracting the area portion of the circle defined by the central angle (160 degrees) from the area of the entire quadrilateral. Area of the circle is $\pi(5^2) = 25\pi$. The ratio of the central angle with respect to 360 degrees is $\frac{145}{360} = \frac{29}{72}$, and it can be applied to the entire area of the circle to find the area portion of the circle defined by the central angle. Thus, the area portion of the circle defined by the central angle is $\frac{25\pi(29)}{72} = \frac{725\pi}{72}$.

Now, we will find the area of the quadrilateral. We can split the quadrilateral into two equal right triangles, so that the central angle becomes 72.5 degrees for each triangle. We can find the area of one of the triangles and then simply multiply the result by 2 to get the area of the quadrilateral. Each tangent of the circle will be the base for the triangle, and the height is the radius of the circle, which is given as 5. The base of the triangle is then: $\tan(72.5^{\circ}) = \frac{b}{5}$, making base $b \approx 15.86$. Area of the triangle is then

$\frac{15.86(5)}{2}$ = 39.65, and the area of the quadrilateral is 2(39.65) = 79.3. Thus, the area of the shaded region is 79.3 – $\frac{725\pi}{72}$ ≈ 48.

Answer: C

182. Find the measure, in radians, of the vertex angle in the triangle formed by the intersection of the two equations and the *y*-axis shown in the figure.

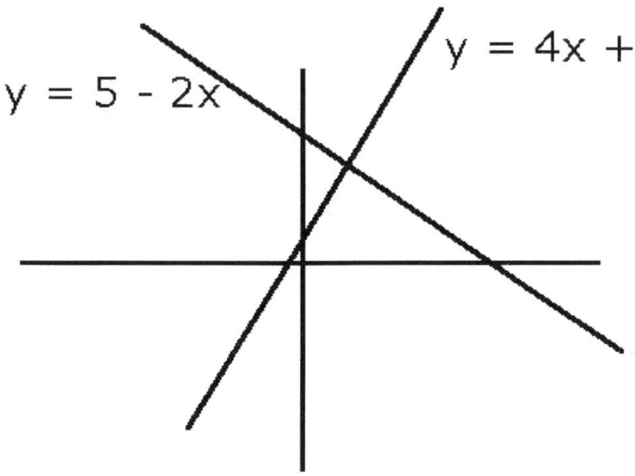

A. 1.5

B. 1

C. 3

D. 2

The *y*-intercepts for both equations are (0, 1) and (0, 5), so that the distance between them is 4 (base of the triangle whose vertex angle we need). Set the two equations equal to find the *x*-value of their intersection: 5 – 2*x* = 4*x* + 1, so that *x* = $\frac{2}{3}$. This is the height of the triangle to the vertex angle (or horizontal distance from the origin to the intersection point). The y-value of the intersection is then $4(\frac{2}{3}) + 1 = \frac{11}{3}$, and the angle between the line of the equation $y = 4x + 1$ and the y-axis is $\tan^{-1}\{\frac{\frac{2}{3}}{\frac{8}{3}}\} \approx 14.04$ degrees.

Angle made by the line of the equation $y = 5 - 2x$ and the y-axis is $\tan^{-1}(\frac{\frac{2}{3}}{\frac{4}{3}}\} \approx 26.57$ degrees. Thus the vertex angle measure we need is $180 - 26.57 - 14.04 \approx 139$ degrees. In radians this is $139(\frac{\pi}{180}) \approx 2$. We could have used the hypotenuses of the two equations and the y-axis segment to find the vertex angle using the *Law of Cosines*, but it would take more calculations to arrive at the same answer.

Answer: D

183. If an equilateral triangle, with side *a*, is inscribed in a circle with radius *r*, expression that defines the radius *r* in terms of the triangle side *a* is

A. $2a\sqrt{3}$

B. $a\sqrt{3}$

C. $\frac{a}{\sqrt{2}}$

D. $\frac{a}{\sqrt{3}}$

It is best to draw a figure to better understand this problem. Since each of the three radiuses bisects each angle of the equilateral triangle, we have six identical 30-60-90 right triangles created.

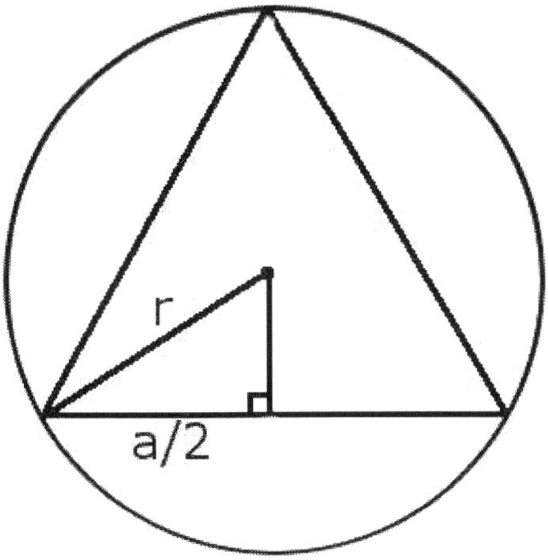

One of the six right triangles is drawn in the figure. The triangle side *a* must be divided in half to be the side of the right triangle. We know that the ratio of sides of a 30-60-90 triangle is 1:$\sqrt{3}$:2. This means that side that equals 1 will be opposite of the 30-degree angle, side that is $\sqrt{3}$ will be opposite of the 60-degree angle, and side that is 2 (the hypotenuse) will be opposite of the 90-degree angle. Since *r* (the hypotenuse) is an opposite side to a 90-degree angle, the smallest side is the vertical side that will be $\frac{r}{2}$ (in terms of *r*), and $\frac{a}{2}$ (the opposite side of the 60-degree angle) must be $(\frac{r}{2})\sqrt{3}$. This means that $\frac{a}{2} = (\frac{r}{2})\sqrt{3}$, so that $r = \frac{a}{\sqrt{3}}$.

Answer: D

184. A random dart will land on the following figure with probability 1. If the radius of the circle that is inscribed in a square is $\frac{1}{9}$ the largest radius, and the mid-size circle is $\frac{1}{5}$ the largest radius, what is the probability that the dart will land on the shaded portion of the figure?

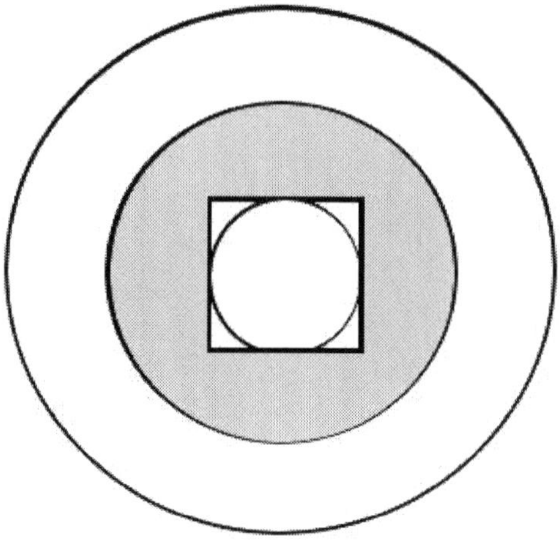

A. $\dfrac{\frac{81\pi}{25} - 4}{81\pi}$

B. $\dfrac{81\pi}{25} - 4$

C. $\dfrac{81\pi}{25} - 1$

D. $\dfrac{\frac{81\pi}{25} - 1}{81\pi}$

If we let the radius of the largest circle be 9, then its area is 81π. The radius of the mid-size circle is $\frac{9}{5}$, so that its area is $\frac{81\pi}{25}$. Since the radius of the smallest circle is 1, its diameter (also the square side measure) is 2, so that the area of the square is 4. Thus, the area of the shaded region is $\frac{81\pi}{25} - 4$. This means the probability that the dart hits the shaded region is $\dfrac{\frac{81\pi}{25} - 4}{81\pi}$, and this is true because the entire probability space consists of the region that is the largest circle area of 81π.

Answer: A

185. If a circle with radius *r* is inscribed in an equilateral triangle with side *s*, function *s*(*r*) is expressed as

A. 12*r*

B. $2\sqrt{3}r$

C. 6*r*

D. cannot be determined

We need to draw a picture to understand the best way to express triangle side *s* in terms of the inscribed circle radius *r*.

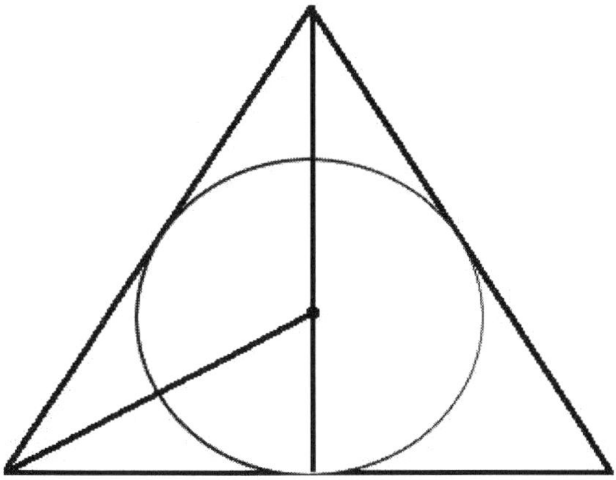

To avoid possible burial with the square root signs, we can work with the area of the isosceles triangle inside the left half of the equilateral triangle. Base of this isosceles triangle is the equilateral triangle side *s*, and height is the radius of the circle *r*. Thus, the area of the isosceles triangle is $\frac{sr}{2}$. Now, we can find the same area of this isosceles triangle by subtracting the area of the small right triangle inside the left half of the equilateral triangle from the left half of the equilateral triangle itself. The small right triangle has a height *r* and base $\frac{s}{2}$. Its area is then $\frac{\frac{s}{2}}{2} = \frac{sr}{4}$. Area of the left half of the equilateral triangle is $\frac{s^2\sqrt{3}}{8}$. Thus, the area of the same isosceles triangle we found earlier is $\frac{s^2\sqrt{3}}{8} - \frac{sr}{4}$. Now we set the two expressions for the isosceles triangle equal: $\frac{s^2\sqrt{3}}{8} - \frac{sr}{4} = \frac{sr}{2}$, which gives $s = 2\sqrt{3}r$.

Answer: B

186. 2,500 feet of cubicle material will be used for construction of eight adjacent offices that make up a rectangular region (see figure). Find the maximum area of the region.

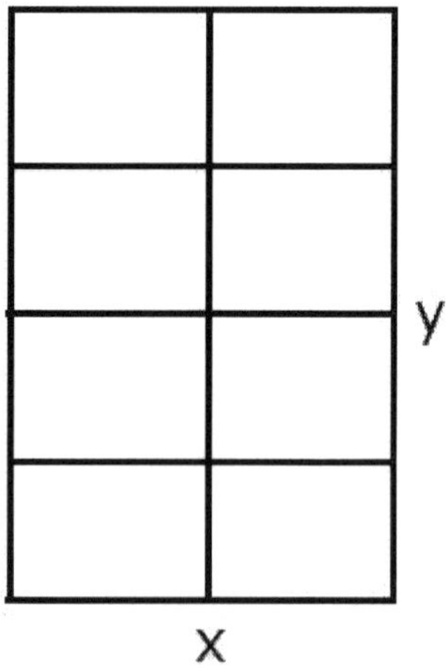

A. 99,875 ft ^{2}

B. 102,181 ft ^{2}

C. 104,167 ft ^{2}

D. 112,346 ft ^{2}

There are five lengths of x (five additional walls parallel to the length of x), and three widths of y (two additional walls parallel to the width y). Thus, we have $2{,}500 = 5x + 3y$, so that $y = \frac{2{,}500 - 5x}{3}$. The area of the region is *xy*, so that $A = \frac{x(2{,}500 - 5x)}{3} = \frac{-5x^2}{3} + \frac{2500x}{3}$. This is a quadratic function. The *x*-value that will produce the maximum area (this function) is determined by the relationships of the constants, $\frac{-b}{2a} = \frac{\frac{-2500}{3}}{2\left(-\frac{5}{3}\right)}$ = 250 feet. Plugging this *x*-value produces the greatest possible area $\frac{250(2500-5(250))}{3} \approx 104{,}167$ ft ^{2}.

Answer: C

187. Find the maximum vertical distance between the graphs of $y = -2x + 5$ and $y = (x + 2)^2 - 5$.

A. 10

B. 15

C. 12

D. 16

Expanding the second equation, we have $y = x^2 + 4x - 1$. The shortest distance will be determined in the interval where the two functions overlap or where there is a closed area bounded by these two graphs. Since the linear equation is above the quadratic equation, we subtract the quadratic equation from the linear one (we subtract the *y*-values of the two equations because the *y*-values determine the vertical distance): $y = -2x + 5 - (x^2 + 4x - 1)$, giving $y = -x^2 - 6x + 6$. Now, since the graph of this composite function opens down, it will have a maximum point (vertex) at some *x*-value, so we find it using the constant formula $\frac{-b}{2a}$ for the *x*-value of the vertex (maximum) point: $x = \frac{-(-6)}{2(-1)} = -3$, which gives the maximum vertical distance $y = -(-3)^2 - 6(-3) + 6 = -9 + 18 + 6 = 15$.

Answer: B

188. A point with polar coordinates $(70, \frac{-3\pi}{4})$ has Cartesian coordinates

A. $(-70\sqrt{2}, 70\sqrt{2})$

B. $(-35\sqrt{2}, 35\sqrt{2})$

C. $(-35\sqrt{2}, -35\sqrt{2})$

D. $(35\sqrt{2}, -35\sqrt{2})$

The radius r is 70, and the angle is $\frac{-3\pi}{4}$, which means that the angle is in quadrant III. Since $x = r\cos\theta$ and $y = r\sin\theta$, then $x = 70\cos(\frac{-3\pi}{4})$ and $y = 70\sin(\frac{-3\pi}{4})$, so that $x = -35\sqrt{2}$ and $y = -35\sqrt{2}$.

Answer: C

189. A parabola-shaped doorway has a height of 12 feet at the center, and is 8 feet wide. If a refrigerator of width 6 feet must be pushed through the doorway, what is the maximum height this refrigerator can have? Assume that the floor and doorway edges are frictionless.

A. 4 ft

B. 4.5 ft

C. 5.25 ft

D. 3 ft

The parabola-shaped doorway width of 6 can be expressed as the horizontal distance from (0, 0) to (8, 0). The vertex x-value of the doorway will be the midpoint of its width, namely (4, 0). Since the height of the doorway at the center is the vertex point, its coordinates are (4, 12). We now know that the parabola can be expressed as $y = a(x - 4)^2 + 12$ using the known values of the vertex. The constant a will tell us how the graph will shrink, because if $|a| < 1$, the parabola tends to get wider and spread out more. Note that this phenomenon will also increase the maximum height for the refrigerator. Since the graph passes through the point (0, 0), we have $0 = a(0 - 4)^2 + 12$, so that $a = -\frac{12}{16} = -\frac{3}{4}$. This means that the parabola is wider compared to the parabola with the same vertex but having $a = -1$. We now have the parabola equation $y = -\frac{3}{4}(x - 4)^2 + 12$.

Now, since the refrigerator is 6 feet wide, its width half is 3 (its center of the width matches the center of the width of the doorway), which means that the x-value of the point of the refrigerator's first vertex determining its height is $4 - 3 = 1$. Plugging in this x-value into the parabolic equation will determine its maximum fitting height: $y = -\frac{3}{4}(1 - 4)^2 + 12 = \frac{21}{4} = 5.25$ feet.

Answer: C

190. In the following right triangle, what is the value of $\dfrac{\left\{\dfrac{\csc(A)}{\sec(B)\cot(A)}\right\}}{\tan(B)}$**?**

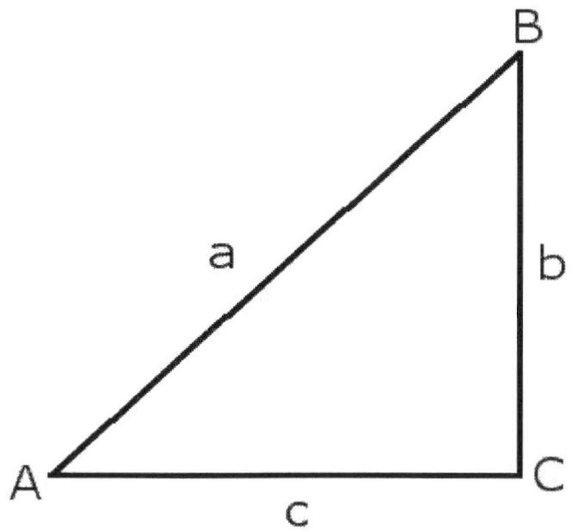

A. 1

B. $\frac{b}{c}$

C. $(\frac{b}{c})^2$

D. $\frac{c}{b}$

We have cscA = $\frac{a}{b}$, and secBcotA = $(\frac{a}{b})(\frac{c}{b}) = \frac{ac}{b^2}$, so that $\frac{\csc(A)}{\sec(B)\cot(A)} = \frac{\frac{a}{b}}{\frac{ac}{b^2}} = \frac{b}{c}$.

TanB = $\frac{c}{b}$, so that $\frac{\left\{\frac{\csc(A)}{\sec(B)\cot(A)}\right\}}{\tan(B)} = \frac{\frac{b}{c}}{\frac{c}{b}} = (\frac{b}{c})^2$.

Answer: C

191. If a line is defined by two parametric equations $x = t - 1$ and $y = -2t + 4$, what is the slope of the line?

A. 2

B. -2

C. -1

D. 1

Taking $t = 0$, we have the point (-1, 4), and with $t = 1$ we have the point (0, 2). The slope of the line is thus $\frac{2-4}{0-(-1)}$ = -2.

Answer: B

192. In the following figure of a cube, the two longest diagonals intersect at point P. Find the measure of angle x at intersection point P.

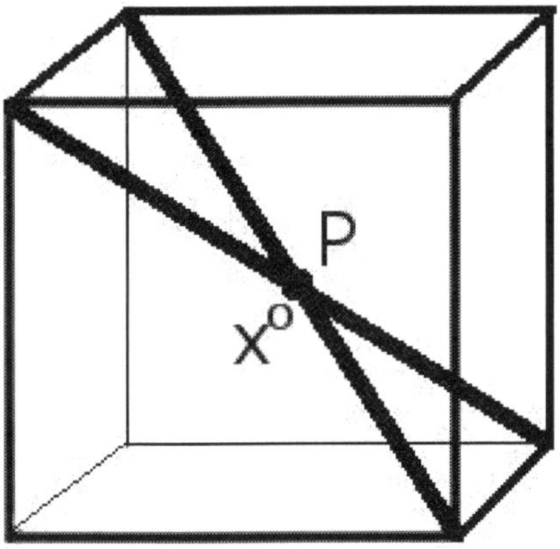

A. 141°

B. 71°

C. 39°

D. 109°

Angle *x* in question is the vertex angle at point P of an isosceles triangle with two diagonal halves as its two equal sides. It is easy to simply use the *Law of Cosines* to find this obtuse angle *x* at point P. If we let each side of the cube be 1, its longest diagonal is $\sqrt{1^2+1^2+1^2} = \sqrt{3}$. Thus, the diagonal half (one of the two equal sides of an isosceles triangle with vertex angle *x* at P) is $\frac{\sqrt{3}}{2}$. The base of this isosceles triangle is $\sqrt{1^2+1^2} = \sqrt{2}$. Thus, we have $(\sqrt{2})^2 = (\frac{\sqrt{3}}{2})^2 + (\frac{\sqrt{3}}{2})^2 - 2(\frac{\sqrt{3}}{2})(\frac{\sqrt{3}}{2})\cos(x)$, so that $x = \cos^{-1}(-\frac{1}{3}) \approx 109°$.

Answer: D

193. If an open right circular cylinder with radius 2 is inscribed in a sphere with radius 7, find the surface area of the cylinder.

A. $12\pi\sqrt{5}$

B. 24π

C. $24\pi\sqrt{5}$

D. $8\pi + 24\pi\sqrt{5}$

We draw a figure to better understand what this problem requires.

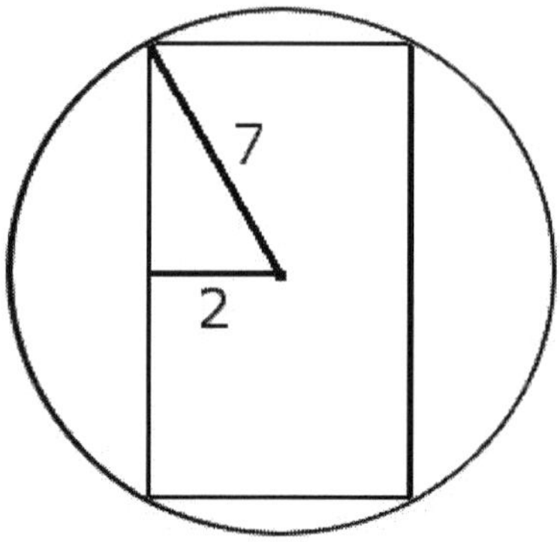

This is a 2-dimensional illustration of a sphere inscribing the open right circular cylinder. Using the Pythagorean Theorem, the height of the triangle drawn is $\sqrt{7^2 - 2^2} = 3\sqrt{5}$, which means that the cylinder height is $2(3\sqrt{5}) = 6\sqrt{5}$. Since the cylinder is open, we need its outer surface area that consists of its height and circle circumference, so it is $2\pi rh = 2\pi(2)(6\sqrt{5}) = 24\pi\sqrt{5}$.

Answer: C

194. If $\sin^{-1}(y) = \arctan(3y)$, then for all $y > 0$, $y =$

A. 0.929

B. 0.785

C. 0.943

D. 0.891

Since $y > 0$, the angle is in quadrant I or II. Since the tangent is positive, it must be precisely in quadrant I. Taking the sine on both sides of the equation we have $\sin(\sin^{-1}(y)) = \sin(\tan^{-1}(3y))$, so that $y = \sin(\tan^{-1}(3y))$. We have $\tan^{-1}(3y)$, which means that $\tan \theta = \frac{3y}{1}$ (opposite over adjacent sides). The third side (hypotenuse) must then be $\sqrt{1 + 9y^2}$. This means that $y = \sin \theta = \frac{3y}{\sqrt{1+9y^2}}$. Squaring both sides and then simplifying, we have $y^2(1 + 9y^2) = 9y^2$, which gives $9y^4 - 8y^2 = 0$. Factoring this, we have $9y^2(y^2 - \frac{8}{9}) = 0$, and the positive root that we need is $y = \sqrt{\frac{8}{9}} \approx 0.943$.

Answer: C

195. If a line segment of an isosceles triangle that has endpoints (-2, -5) and (3, -4) is a base of the triangle, the third possible point for the triangle vertex could be

A. (-1, 7)

B. (0, 2)

C. (2, -12)

D. (2, 8)

The slope of the line through the two given points is $\frac{-4+5}{3+2} = \frac{1}{5}$. This means that the height from the midpoint of the base to the vertex of the triangle will have a slope -5 (negative reciprocal of the original slope). The midpoint of the base has coordinates $(\frac{3-2}{2}, \frac{-4-5}{2}) = (\frac{1}{2}, -\frac{9}{2})$. Equation of the line through the vertex and the midpoint of the base is $y = -5(x - \frac{1}{2}) - \frac{9}{2}$, that is, $y = -5x - 2$. Plugging in the choices in this equation, we can see that only C works.

Answer: C

196. In an oblique triangle ABC, $_m$<A = 36°, opposite side to <A is 3. Also, $_m$<B = 47°, and the opposite side to it is *x*. Suppose that this triangle became a right triangle with the two mentioned sides being the same as before. What would the third side be equal to?

A. 6

B. 4.8

C. 4

D. 7

Using the Law of Sines, we have $\frac{\sin(36^o)}{3} = \frac{\sin(47^o)}{x}$, giving $x \approx 3.73$. If this were a right triangle, the third side would have been $\sqrt{3^2 + 3.73^2} \approx 4.8$.

Answer: B

197. Find the directrix for the graph of $x = -\sqrt{14y}$.

A. $x = \frac{7}{2}$

B. $x = -\frac{7}{2}$

C. $y = \frac{7}{2}$

D. $y = -\frac{7}{2}$

Squaring both sides, we have $x^2 = 14y$. This is an equation of a parabola, with strictly values for $y \geq 0$. This parabola thus opens up, and is in standard position with vertex point (0, 0). Since the equation of a parabola in standard form is $x^2 = 4py$ (with *p* being the same distance from the vertex to the focus and from the vertex to the directrix line), with directrix being the line $y = -p$, we have $4p = 14$, so that $p = \frac{7}{2}$. The directrix for this parabola is thus the line $y = -\frac{7}{2}$.

Answer: D

198. The following figure shows a square inscribed in a right triangle. The base of a triangle is $\frac{3}{7}$ the height. If the area of the triangle is 70, find the area of the inscribed square (nearest whole).

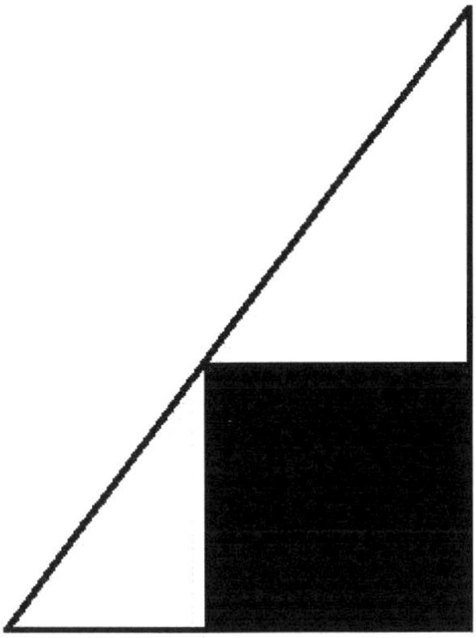

First we find the dimensions of the triangle in question. If the height of the triangle is h, and base b, then $b = \frac{3h}{7}$. The area is $70 = h(\frac{3h}{7})$, so that $h = \sqrt{\frac{490}{3}}$ and $b = \frac{3\sqrt{\frac{490}{3}}}{7}$. If we let the side of the square be x, we can combine the areas of three regions that make up the triangle. We have $70 = \frac{3\sqrt{\frac{490}{3}}}{7} \cdot \frac{x}{2} + x^2 + x(\sqrt{\frac{490}{3}} - x)$, so that $x = (\frac{980}{17})(\sqrt{\frac{3}{490}})$.

The area of the square is thus $[(\frac{980}{17})(\sqrt{\frac{3}{490}})]^2 \approx 20$.

Answer: 20

199. A point with Cartesian coordinates $(-\frac{17}{2}, -\frac{17\sqrt{3}}{2})$ has polar coordinates

A. $(-17, \frac{4\pi}{3})$

B. $(-17, \frac{\pi}{3})$

C. $(17, -\frac{\pi}{3})$

D. $(17, -\frac{2\pi}{3})$

Since the polar coordinates exist in the form of (r, θ), and since $x^2 + y^2 = r^2$, then $r^2 = (-\frac{17}{2})^2 + (-\frac{17\sqrt{3}}{2})^2 = \frac{289}{4} + \frac{867}{4} = \frac{1,156}{4}$, so that $r = 17$. Since the angle θ is in quadrant III (both x and y are negative), we can use either the arcsine or arccosine to find the angle θ: $\sin^{-1}(\frac{y}{r}) = \sin^{-1}\{-\frac{\frac{17\sqrt{3}}{2}}{17}\} = \sin^{-1}(-\frac{\sqrt{3}}{2})$ = -60 degrees (this is what your calculator will give). However, the angle we need is in quadrant III and we know that sine of θ is also equal to the value $-\frac{\sqrt{3}}{2}$ for $\theta = -60 - 180 + 2(60) = -120$, or $(-120)(\frac{\pi}{180}) = -\frac{2\pi}{3}$ in radians.

Answer: D

200. The following figure shows a square circumscribing a circle with radius *r* that inscribes a rectangle with length *x* and width *y*. Express the perimeter of the square as a function of the width of the rectangle if the area of the rectangle is 40.

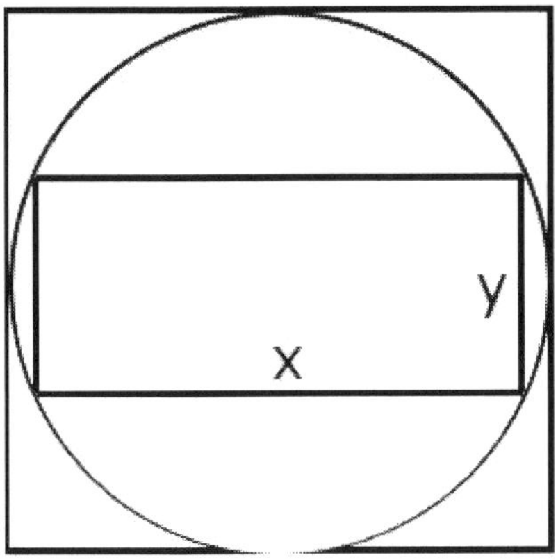

A. $\frac{4\sqrt{y^2 + 1{,}600}}{y}$

B. $\frac{4\sqrt{y^4 + 1{,}600}}{y}$

C. $4\sqrt{y^2 + 1{,}600}$

D. $4\sqrt{y^4 + 1{,}600}$

Since the diagonal of the rectangle passes through the center of the circle, the diameter of the circle (which is also the side of the square circumscribing the circle) is $\sqrt{x^2 + y^2}$. The perimeter of the square is then $4\sqrt{x^2 + y^2}$. Since $xy = 40$, $x = \frac{40}{y}$. The perimeter of the square then becomes $4\sqrt{(\frac{40}{y})^2 + y^2} = 4\sqrt{\frac{y^4 + 1{,}600}{y^2}} = \frac{4\sqrt{y^4 + 1{,}600}}{y}$.

Answer: B

201. If a bicycle's sprocket makes 50 revolutions per minute, find the speed of the bicycle, in miles per hour (nearest whole), if its wheel radius is 3 feet.

A. 10

B. 11

C. 12

D. 13

The circumference of the bicycle wheel is $2\pi(3) = 6\pi$ feet. The bicycle covers this distance as its wheel completes one revolution, so that the distance covered per minute is $6\pi(50) = 300\pi$ feet. Thus the speed of the bicycle is (300π feet / minute)(60 minutes / hour)(mile / 5,280 feet) ≈ 11 miles per hour.

Answer: B

202. If the larger of the two similar equilateral triangles inscribes a circle, and the triangle area ratio is 12:1, find the area of the circle if the side of the small triangle is 1.

A. $\frac{5\pi}{2}$

B. $\frac{\pi}{3}$

C. π

D. $\frac{\pi}{4}$

Since the triangle area ratio is 12:1, the side ratio must then be $\sqrt{12}$:1 (because the ratio of the areas of two similar triangles is the square of the ratio of their sides). Since the side of the smaller triangle is 1, the larger triangle side is $\sqrt{12}$. Now, the inscribed circle in the larger triangle will have a radius *r* that is equal to the height of one of the three equal isosceles triangles that the equilateral triangle can be divided into (the three isosceles triangles will share the same vertex point that is the center of the inscribed circle). The base of the isosceles triangle is the side of the equilateral triangle, that is, $\sqrt{12}$. The area of this isosceles triangle is $\frac{\sqrt{12}r}{2}$. Since there are three isosceles triangles that make up this larger equilateral triangle, the area of one of the isosceles triangles is $\frac{(\sqrt{12})^2\left(\frac{\sqrt{3}}{4}\right)}{3} = \sqrt{3}$. Combining the two expressions for the isosceles triangle area, we have $\frac{\sqrt{12}r}{2} = \sqrt{3}$, so that $r = 1$. Area of the circle is then simply π.

Answer: C

203. If a car travels along a line *y* = 4*x* – 5 at a rate of 25 feet per second from the starting point at *x* = -27, find the number of milliseconds (nearest whole) that pass until the car hits the *x*-value of 10 feet.

A. 610

B. 6

C. 6,102

D. 931

The coordinate of the starting point is (-27, -113). The final destination coordinate is (10, 35). The distance between these two points is $\sqrt{(-27-10)^2+(-113-35)^2} = \sqrt{23{,}273}$ feet. The time, in milliseconds, that it takes the car to cover this distance is ($\sqrt{23{,}273}$ feet)(second / 25 feet)(1000 milliseconds / second) ≈ 6,102.

Answer: C

204. Jose decides to go on a dangerous climbing trip to the Everest. He experiences changing air pressure 1,000 feet from the ground level, which is 14.2 pounds per square inch. At the altitude of 2,000 feet Jose experiences a drop in

air pressure by $\frac{1}{2}$ pounds per square inch. What model best shows the linear relationship between air pressure *p* and altitude *a*?

A. $p = -\frac{1}{2{,}000}a - 14.2$

B. $p = 2{,}000a + 14.7$

C. $p = -\frac{1}{2{,}000}a + 14.7$

D. $p = -\frac{1}{1{,}000}a + 14.7$

We can think of the independent variable *a* as the rising altitude, and *p* (the dependent variable) as the decreasing air pressure. The slope of the line determining the relationship between air pressure and altitude is $\frac{-\frac{1}{2}}{2000 - 1000} = -\frac{1}{2{,}000}$. Note that the change in *p* (drop of air pressure by $\frac{1}{2}$ which makes it negative) was given in the problem as the altitude of 2,000 feet from the 1,000 feet was reached by Jose. To complete the model, we find the *p*-intercept *b* or the initial pressure at the ground level (when $a = 0$). We use the known point for this and the slope we already know: $14.2 = -\frac{1}{2{,}000}(1000) + b$, so that the *p*-intercept $b = 14.7$ feet. Thus the equation modeling the air pressure *p* with respect to changing altitude *a* is $p = -\frac{1}{2{,}000}a + 14.7$.

Answer: C

205. A person casts a shadow of length *L* when the angle of the sun above the horizon is *x*, and the shadow length is modeled by $L = \frac{h\sin\left(\frac{\pi}{2} - x\right)}{\sin(x)}$, where *h* is the height of any standing person. If two people stand next to each other with respective heights 6 feet and 5.5 feet, find the difference of the lengths, in inches (nearest 1), of their shadows when the sun is 60 degrees above the horizon.

A. 1

B. 2

C. 3

D. 4

We first simplify the length model. Since $\sin(\frac{\pi}{2} - x) = \cos(x)$, $L = h\cot(x)$. Using $h = 6$ and $x = 60^\circ$ or $\frac{\pi}{3}$ (make sure you are in the correct calculator mode if you choose degrees or radians because the result will not be the same), we have $L_1 = 6\cot(60^\circ) = 6(\frac{\sqrt{3}}{3}) = 2\sqrt{3}$ feet, and $L_2 = 5.5\cot(60^\circ) = \frac{5.5\sqrt{3}}{3}$ feet. The difference between their shadows is thus $|\frac{5.5\sqrt{3}}{3} - \frac{6\sqrt{3}}{3}|$ feet. In inches this is $12|\frac{5.5\sqrt{3}}{3} - \frac{6\sqrt{3}}{3}| \approx 3$ inches.

Answer: C

206. The following figure shows the isosceles triangle with the sides that each measure 12 and angle ɵ between them. The area of the triangle in terms of angle ɵ can be expressed as

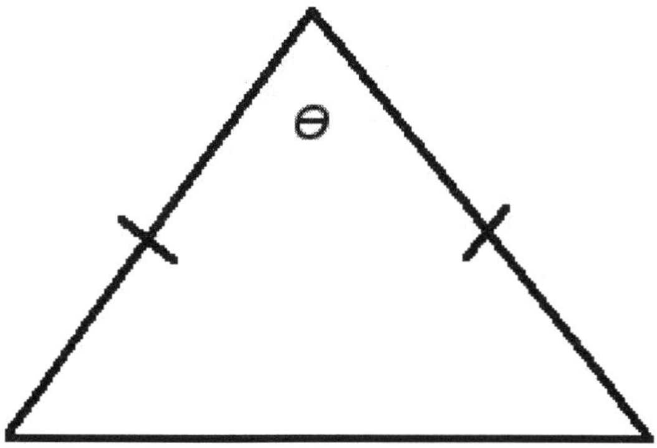

A. 144cos2ɵ

B. $72\sin(\frac{\theta}{2})$

C. $72\sin\theta$

D. $72\tan\theta$

First, we split the triangle in half, so that we will be dealing with half of the angle θ for each of the two identical right triangles created, or $\frac{\theta}{2}$. Let's work with only one of the two right triangles we created. If we let the base of the right triangle be x, and the height y, then $\sin(\frac{\theta}{2}) = \frac{x}{12}$, so that x (half of the base of the original isosceles triangle) = $12\sin(\frac{\theta}{2})$, so that $2x$ (base for the isosceles triangle) = $24\sin(\frac{\theta}{2})$. Now, using the Pythagorean Theorem, the height for the right triangle (and isosceles triangle) is $y = \sqrt{12^2 - x^2} = \sqrt{144 - 144\sin^2(\frac{\theta}{2})} = \sqrt{144(1 - \sin^2(\frac{\theta}{2}))} = \sqrt{144\cos^2(\frac{\theta}{2})} = 12\cos(\frac{\theta}{2})$. Thus the area of the isosceles triangle is $\frac{xy}{2} = \frac{\left(24\sin\left(\frac{\theta}{2}\right)\right)\left(12\cos\left(\frac{\theta}{2}\right)\right)}{2} = 144\sin(\frac{\theta}{2})\cos(\frac{\theta}{2})$.
Notice that this expression resembles $2\sin x\cos x$, and since $2\sin x\cos x = \sin 2x$, we have the area of this triangle as $144\sin(\frac{\theta}{2})\cos(\frac{\theta}{2}) = (72 \cdot 2)\sin(\frac{\theta}{2})\cos(\frac{\theta}{2}) = 72\sin(\frac{2\theta}{2}) = 72\sin\theta$.

Answer: C

207. Using the same figure, find value of θ, in radians, that maximizes the area of the triangle.

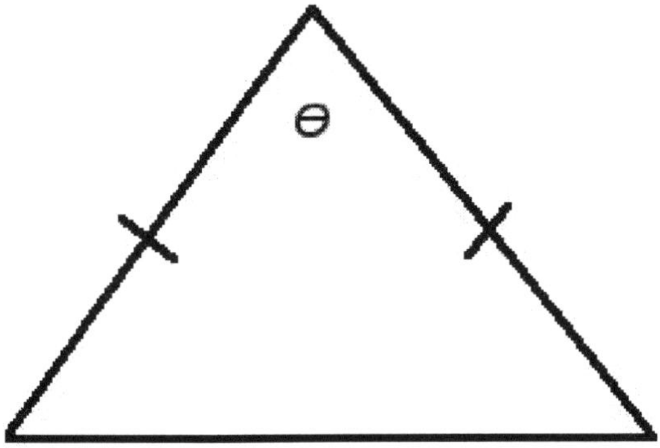

A. $\frac{\pi}{2}$

B. $\frac{\pi}{2} + 2n\pi$, n an integer

C. $\frac{\pi}{2} \pm 2n\pi$, n an integer

D. $\frac{\pi}{4}$

Since the area of the triangle is 72sinθ, we know that sinθ has a maximum value of 1 when $\theta = \frac{\pi}{2}$. Thus, the maximum area for the triangle is 72 when $\theta = \frac{\pi}{2}$. Note that in the context of a triangle choice B is wrong, because any triangle cannot have angles greater than or equal to 180 degrees or π radians. Choice C is wrong for the same reason, and triangles cannot have negative angle values.

Answer: A

208. The following figure shows two identical circles that are tangent to each other, with tangent lines to the circles meeting at the point P. If the angle between

the tangent lines is 40°, and radius of each of the two circles is 4 inches, find the area, in square feet (nearest tenth), of the shaded region.

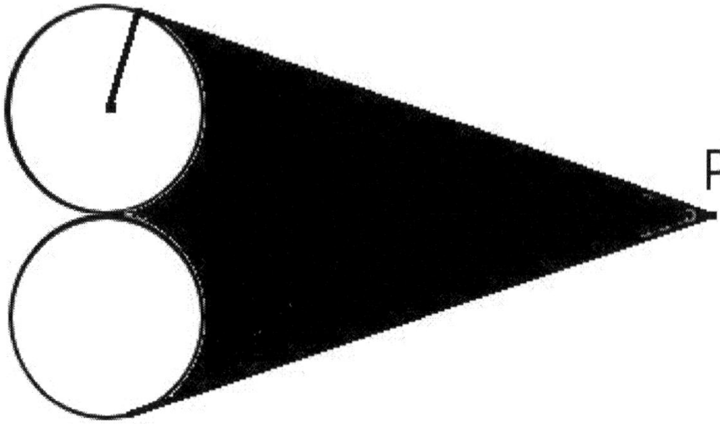

A. 1.11 ft^2

B. 136.8 ft^2

C. 0.95 ft^2

D. 159.1 ft^2

The best way to find the area of the shaded region is to subtract the area portions of the circles taken up by the radiuses and tangent lines from the areas of the triangles that include the point P and radiuses of the circles. First, note that the imaginary tangent line that is exactly between the two tangent lines is equal to the tangent lines because it starts from the point P as well and is tangent to the two circles at the point common to them (the point that makes them adjacent). This line cuts the shaded region in half, and one of these two halves contains two identical right triangles (with base as the radius of the circle for each of the two triangles). Notice that there are four such right triangles that extend the shaded region into a triangle by using the area portions of the two circles. Take a look at the figure below.

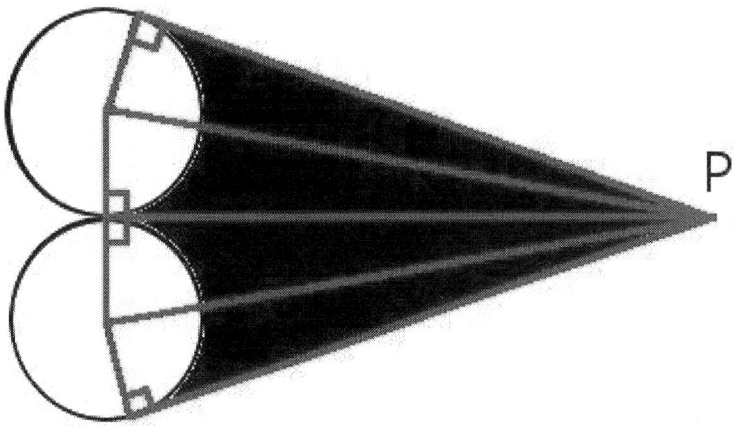

We only need to find the area portion of one circle taken up by one of these four right triangles and then subtract that from the area of one of these four right triangles (we simply multiply the end result by 4 to get the area of the entire shaded region).

The angle of the entire shaded region that was at point P is now $\frac{1}{4}$ its value, namely $\frac{40}{4}$ = 10° as the smallest angle in each of the four identical right triangles. We can simply work with the top right triangle whose height is the top tangent line. Since the radius is 4, the height h is calculated: $\tan(10°) = \frac{4}{h}$, giving h = 22.68512728. The area of the triangle is $\frac{4(22.68512728)}{2}$ = 45.37025456. Now, to find the area of the circle taken up by this triangle, we need to find the smallest central angle of the circle that is also the third angle of the small right triangle (angle whose opposite side is the upper tangent line to the top circle). This third angle (and smallest central angle of the circle) in the triangle is 180 – 90 – 10 = 80°, and since the area of the circle is 16π, the ratio of the central angle to 360 degrees is $\frac{80}{360} = \frac{2}{9}$, so that the area portion of the circle taken up by the triangle is $(\frac{2}{9}) \cdot 16\pi = \frac{32\pi}{9}$. The area of the shaded region part of one of the four triangles is then $45.37025456 - \frac{32\pi}{9}$. Thus the area of the entire shaded region is simply four times this amount, namely $4(45.37025456 - \frac{32\pi}{9})$, which is in square inches. In square feet this is $\frac{4\left(45.37025456 - \frac{32\pi}{9}\right)}{144} \approx 0.95$.

Answer: C

209. The following figure shows two lines in the standard *xy*-plane with the given *y*-intercepts and positive slopes. The angle between the two lines can be expressed as

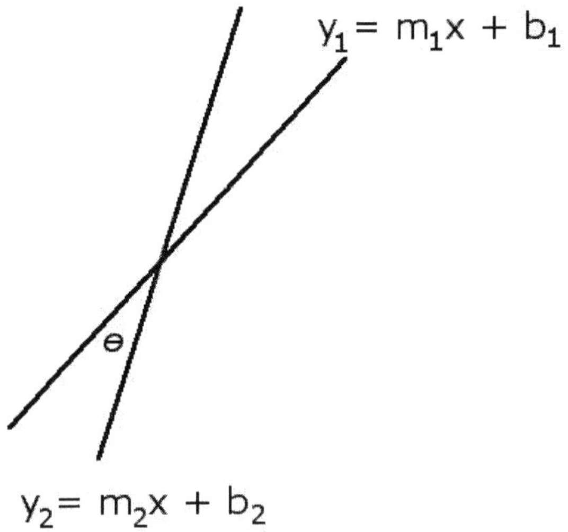

A. $\tan^{-1}(m_1) - \tan^{-1}(m_2)$

B. $\cos^{-1}(m_1) - \sin^{-1}(m_2)$

C. $\tan^{-1}(m_2) - \tan^{-1}(m_1)$

D. $\sin^{-1}(m_2) - \cos^{-1}(m_1)$

We can construct a figure showing a triangle inside a larger triangle, with both triangles' hypotenuses pertaining to the two slope lines shown in the figure. Take a look below.

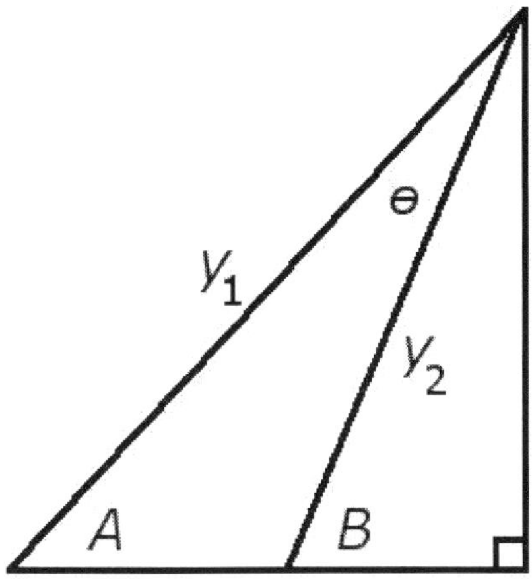

Using the labels for angles A and B, we can see that $\tan A = m_1$ (change of y divided by the change of x) and $\tan B = m_2$, which means that $A = \tan^{-1}(m_1)$ and $B = \tan^{-1}(m_2)$. Since $A + \theta = B$, $\theta = B - A = \tan^{-1}(m_2) - \tan^{-1}(m_1)$.

Answer: C

210. Using the formula found in the previous example, find the angle formed in degrees (the nearest tenth) between the lines $y_1 = 2x$ and $y_2 = 5x - 6$.

A. 15.9°

B. 14.7°

C. 16.2°

D. 15.3°

Using the formula we found, we have $\theta = \tan^{-1}(5) - \tan^{-1}(2) \approx 15.3°$.

Answer: D

211. A line whose equation is $y = kx + b$ is perpendicular to a curve whose equation is $y = x^2 + 4$ at the point $(4, y)$. Find b.

A. 20

B. 21

C. 20.5

D. 19.5

Even though the curve is a not a straight line, it has infinite instant slopes (slopes of the tangent lines to it) at different points (x, y). For example, the curve has a slope of 0 at $x = 0$, because this point is the breaking point where the function starts to increase (where the slope becomes positive). We are asked to find the slope of the curve when $x = 4$. Using the formal definition of a derivative (instant slope) and $f(x) = x^2 + 4$, we have the slope formula as $\lim_{h \to 0} \frac{f(a+h) - f(a)}{h}$ where $a = 4$. Note that this definition makes sense with the formal definition of a slope of a straight line too, because the numerator of the formula shows the change in y-values, and the denominator of the formula shows the change in x-values. We need the change of the x-values to be approaching 0 (because we want the instant slope at the particular point), which is why $h \to 0$. Plugging the function $f(x) = x^2 + 4$ we have into the derivative (instant slope) formula, we have $\lim_{h \to 0} \frac{(a+h)^2 + 4 - (a^2 + 4)}{h}$, so that the slope becomes $\lim_{h \to 0} \frac{2ah + h^2}{h}$. Since we need to find the slope of the curve when $x = a = 4$, we have $\lim_{h \to 0} \frac{8h + h^2}{h}$, so that it is simplified to $\lim_{h \to 0} (8 + h)$, making the slope at $x = 4$ as $h \to 0$ become $8 + 0 = 8$, so that the slope of the perpendicular line to the curve at that point $x = 4$ is $-\frac{1}{8}$ (negative reciprocal). The y-coordinate at point $x = 4$ is $y = 4^2 + 4 = 20$, so that $20 = kx + b$ for the straight line. Since the slope of the perpendicular line is $-\frac{1}{8}$, $k = -\frac{1}{8}$. This means that $20 = -\frac{1}{8}(4) + b$, making $b = 20 + \frac{4}{8} = 20 + \frac{1}{2} = 20.5$.

Answer: C

212. A triangle is drawn in a standard *xy*-coordinate plane. Find the area of this triangle if its vertices are at (4, -6), (-3, -2) and (2, -1). Express your answer in the simplest fraction form.

From the picture we draw, the base of this triangle must be a line segment with the endpoints (-3, -2) and (4, -6). The length of this segment is $\sqrt{(-3-4)^2+\left(-2-(-6)\right)^2}$ $= \sqrt{65}$. This is what the base of the triangle is equal to. The slope of the line collinear with this line segment is $\frac{-2-(-6)}{-3-4} = -\frac{4}{7}$. Equation of the line collinear with the base is $y = -\frac{4}{7}(x+3) - 2$, or $y = -\frac{4}{7}x - \frac{26}{7}$. The fastest way to find the height length is to use the distance formula from a point (vertex in this case) to a line (line collinear with the base) if you can remember it. Distance from a line $\boldsymbol{ax + by + c = 0}$ to the point (x_o, y_o) is $\frac{|ax_o + by_o + c|}{\sqrt{a^2+b^2}}$. Note the absolute value symbol in the numerator, because the distance should always be positive (it is possible to have a negative result inside the symbol). The equation of the line we need is $y = -\frac{4}{7}x - \frac{26}{7}$, or in standard form as $-\frac{4}{7}x - y - \frac{26}{7} = 0$. Thus the distance (height length) from (2, -1) to this line is

$$\frac{|-(\frac{4}{7})(2) + (-1)(-1) + (-\frac{26}{7})|}{\sqrt{(-\frac{4}{7})^2 + (-1)^2}} = \frac{27}{\sqrt{65}}.$$

The longer way to find the height (if you forget the distance formula from a point to a line) is to first realize that the line collinear with the height must be perpendicular to the line collinear with the base, making its slope $\frac{7}{4}$. Using the vertex point (2, -1) we find the equation of the line collinear with the height: $y = \frac{7}{4}(x-2) - 1$, or $y = \frac{7}{4}x - \frac{9}{2}$. The point at which the two equations meet is the point we need with the vertex point (2, -1) in order calculate the length of the height. We set the two equations equal: $\frac{7}{4}x - \frac{9}{2} = -\frac{4}{7}x - \frac{26}{7}$, so that $x = \frac{22}{65}$ and $y = (\frac{7}{4})(\frac{22}{65}) - \frac{9}{2} = -\frac{254}{65}$. Thus the point of intersection of the two equations is $(\frac{22}{65}, -\frac{254}{65})$. Thus, the height length is the distance between this point and the vertex at (2, -1). This is $\sqrt{(\frac{22}{65} - 2)^2 + (-\frac{254}{65} + 1)^2} = \frac{27\sqrt{65}}{65}$. Thus, the area of the triangle is $\frac{bh}{2} = \frac{\sqrt{65}\left(\frac{27}{\sqrt{65}}\right)}{2} = \frac{27}{2}$.

Answer: $\frac{27}{2}$

213. In the following figure, the circle diameter is equal to the height of the triangle. If sec(ɵ) = 2sin(ɵ) for some angle ɵ, and hypotenuse of the triangle measures 6, find the area of the figure.

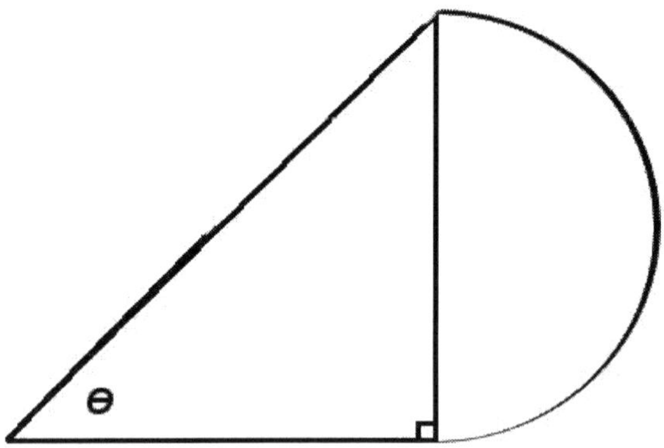

A. $\frac{9\pi}{2}$

B. $\frac{9\pi}{4}$

C. $9 + \frac{9\pi}{2}$

D. $9 + \frac{9\pi}{4}$

If we let a be the base of the tringle, and b the height, using the definition given in this problem, sec(θ) = 2sin(θ), we have $\frac{6}{a} = \frac{2b}{6}$, so that $2ab = 36$, making $b = \frac{18}{a}$. Using the Pythagorean Theorem, we have $6^2 = a^2 + \frac{18^2}{a^2}$, so that $a^4 - 36a^2 + 324 = 0$. This means that $(a^2 - 18)^2 = 0$, giving $a = 3\sqrt{2}$. This makes $b = \frac{18}{3\sqrt{2}} = 3\sqrt{2}$.

Another way to find this result is to see that sec(θ) = 2sin(θ) means that $\frac{1}{\cos(\theta)}$ = 2sin(θ), so that 1 = 2sinθcosθ = sin(2θ), so that 2θ = $\sin^{-1}(1) = 90^\circ$, making θ = 45°. Then, $\sin(45^\circ) = \frac{b}{6}$, so that $\frac{\sqrt{2}}{2} = \frac{b}{6}$, making $b = 3\sqrt{2}$, and the same result for a. Area of the triangle is then $\frac{(3\sqrt{2})^2}{2} = 9$, and the area of the semi-circle is $\pi\frac{\left(\frac{3\sqrt{2}}{2}\right)^2}{2} = \frac{9\pi}{4}$. Thus the area of the figure is $9 + \frac{9\pi}{4}$.

Answer: D

214. If 3sin(2*b*) = $(\frac{1}{4})$cos(2*b*), find the value of b^2. Round your answer to the nearest tenth.

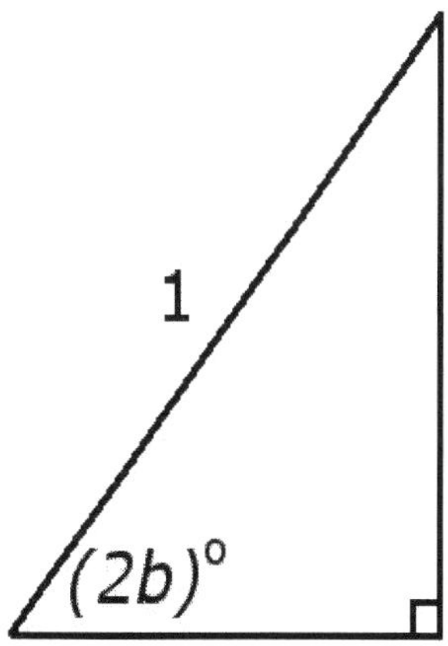

We have 12sin(2*b*) = cos(2*b*). Letting *x* be the base of the triangle, *y* be the height, we have 12*y* = *x*. Using the Pythagorean Theorem, we have $1 = 144y^2 + y^2$, so that $y = \frac{1}{\sqrt{145}}$. Since $\sin(2b) = \frac{1}{\sqrt{145}}$, $2b = \sin^{-1}(\frac{1}{\sqrt{145}})$, making $b^2 \approx 5.7$.

Answer: 5.7

215. In the following figure, an isosceles ΔADE is inscribed in a rectangle ABCD. The rectangle width is half its length. FG || AD. If the perimeter of the rectangle is 26, find the area of ΔEFG.

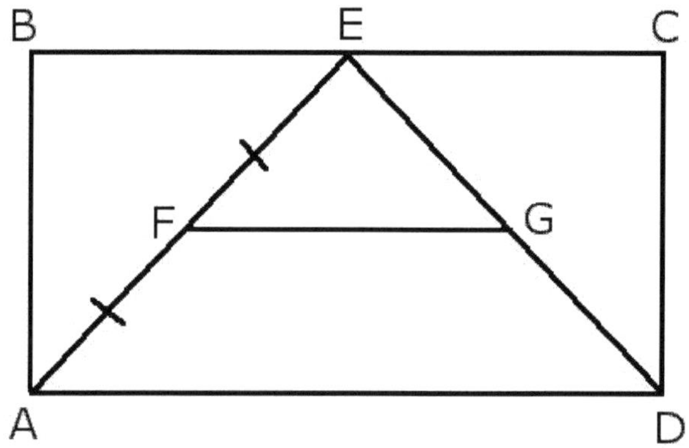

Let *x* be the rectangle length. The width is then $\frac{x}{2}$. Hence 26 = 2*x* + *x* = 3*x*, so that $x = \frac{26}{3}$ (length), and width is $\frac{13}{3}$. Notice that the width equals in measure to the height of the isosceles triangle AED. Since it is isosceles, the height to E will be drawn from the midpoint of AD, and $\frac{AD}{2} = \frac{13}{3}$. Then AE = $\sqrt{2\left(\frac{13}{3}\right)^2} = \frac{13\sqrt{2}}{3}$. Since AF = FE, FE = $\frac{\frac{13\sqrt{2}}{3}}{2} = \frac{13\sqrt{2}}{6}$. Since FE = AF and FG || AD, triangles EFG and ADE are similar (triangle EFG is isosceles). This also means that the height *h* of triangle EFG is half the height H of triangle ADE. Since they are similar, $\frac{FE}{AE}$ = FG/AD, so that $\frac{\frac{13\sqrt{2}}{6}}{\frac{13\sqrt{2}}{3}} = \frac{FG}{\frac{26}{3}}$ making FG = $\frac{13}{3}$. Height h is $\frac{AB}{2}$, or $\frac{13}{6}$. Thus the area of ΔEFG is $\frac{\left(\frac{13}{3}\right)\left(\frac{13}{6}\right)}{2} = \frac{169}{36}$.

Another way (perhaps a faster way) to do this would be the following: after finding AE, it is easy to see that $_{m}$<EAD = $_{m}$<EFG = 45°, because the width and half the length segments of the rectangle are equal (components of hypotenuse AE of ΔADE) and FG || AD. Since $_{m}$<EDA = $_{m}$<EGF = 45° (for the same reasons), $_{m}$<FEG = 90°. Since this makes ΔEFG a right triangle, and EF = EG (isosceles property), FG is its hypotenuse, FE is its base, and EG is its height. Since FE = $\frac{13\sqrt{2}}{6}$ (as was found earlier), the area of the triangle is $\frac{\left(\frac{13\sqrt{2}}{6}\right)\left(\frac{13\sqrt{2}}{6}\right)}{2} = \frac{169}{36}$.

Answer: $\frac{169}{36}$

216. In the following figure, AB = 6, EB = $\frac{AB}{2}$, DC = $\frac{AB}{3}$, DE = $\frac{EC}{2}$, and the slope of the line collinear with segment DC is undefined. Find the slopes of the lines collinear with segments BC and BD. **Note: figure not drawn to scale.**

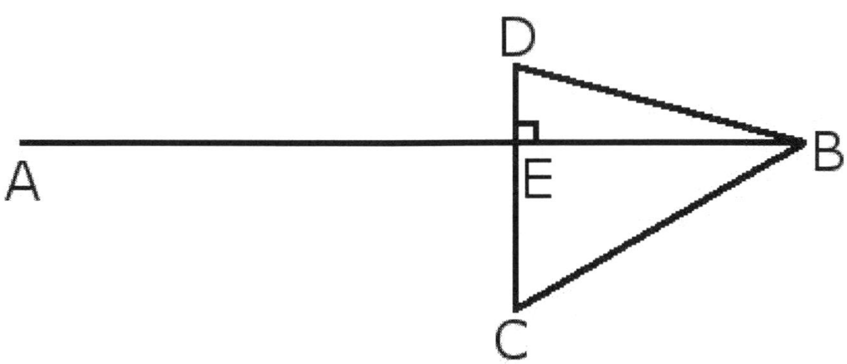

A. $\frac{1}{3}, -\frac{2}{3}$

B. $\frac{2}{9}, -\frac{4}{9}$

C. $-\frac{2}{9}, \frac{4}{9}$

D. $-\frac{1}{3}, \frac{2}{3}$

From the given information, the slope of the line collinear with segment AB must be 0 (since it is perpendicular to DC). EB = $\frac{AB}{2} = \frac{6}{2}$ = 3. DC = $\frac{6}{3}$ = 2. Now, 2DE + DE = 2, so that DE = $\frac{2}{3}$, and EC = $\frac{4}{3}$. Thus, the slope of the line collinear with segment BD is $-\frac{DE}{EB}$ = $-\frac{\frac{2}{3}}{3} = -\frac{2}{9}$ (DB slope line falls from left to right), and the slope of the line collinear with segment BC is $\frac{EC}{EB} = \frac{\frac{4}{3}}{3} = \frac{4}{9}$ (BC slope line rises from left to right).

Answer: C

217. A circle is enclosed in an octagon. If the octagon side is 5, find the area of the region inside the octagon but outside the circle. Round your answer to the nearest whole.

Each interior angle of the octagon measures $\frac{180(8-2)}{8}$ = 135 degrees. The octagon can be divided into eight equal quadrilaterals (kites), each having angles 135, 90, 90, and 360 – 90 – 90 – 135 = 45 degrees. The 45-degree angle is the central angle of each quadrilateral (note that there are eight such central angles, and their sum is 360). See the accompanying figure below.

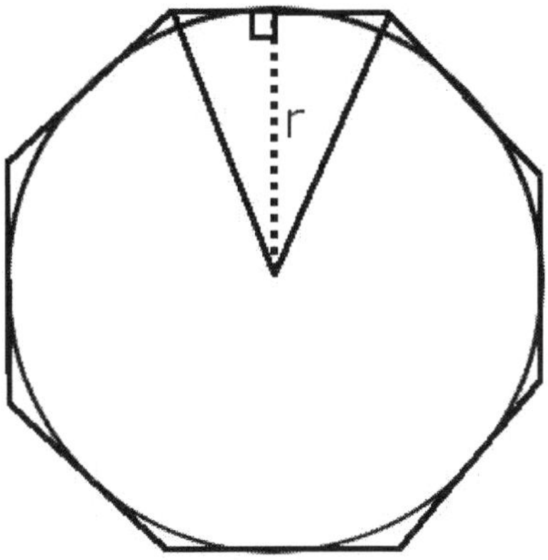

One of these quadrilaterals can be divided in half so that we have a right triangle which consists of the circle radius, half of the octagon side, and line from the circle center to the corner of the octagon. The angles in this right triangle now are $\frac{135}{2}$ = 67.5° (opposite side to this angle is the radius), 90, and $\frac{45}{2}$ = 22.5°. We can use of the two angles (22.5° or 67.5°) inside this triangle to find the radius *r*: tan(22.5°) = $\frac{\frac{5}{2}}{r}$, so that *r* = 6.035533906. Alternatively, we have tan(67.5°) = $\frac{r}{\frac{5}{2}}$, so that *r* = 6.035533906. Circle area is then π(6.035533906^2) = 114.440899. If you do not remember the octagon area formula, you can see that the octagon is divided into eight equal isosceles triangles, each with height that is the radius of the circle and base that is the octagon side, so that the area of the octagon (consisting of combined area of eight isosceles triangles) is $\frac{8\left((5)(6.035533906)\right)}{2}$ = 120.7106781. Area of the region inside the octagon but outside the enclosed circle is 120.7106781 – 114.440899 ≈ 6.

Answer: 6

218. **In the following right △ABC, AE is a median. If the area of △ABC is 50, CD = CE, and AB = $\frac{AC}{3}$, find the area of △AED (nearest whole). Figure not drawn to scale.**

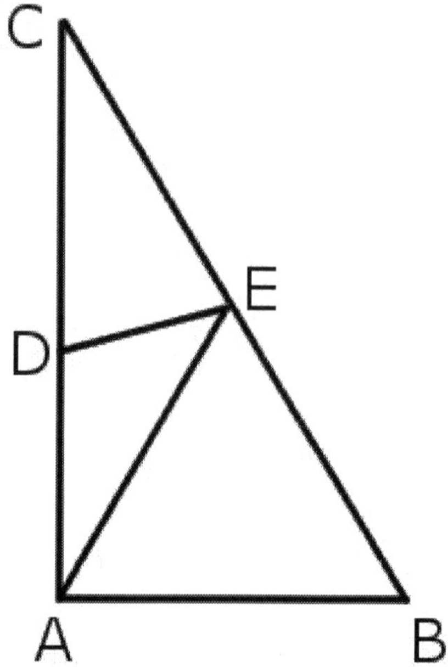

We need to find AE (base of ΔAED) and height from AE to D to find the area of ΔAED. Using the information given, we have $AC^2 = 150$, so that AC = 12.24744871. This makes AB = 4.082482905. BC = $\sqrt{(4.082482905)^2 + (12.24744871)^2}$ = 12.90994448. Since AE is a median, CE = $\frac{BC}{2} = \frac{12.90994448}{2}$ = 6.454972242. We can use the *Law of Sines* to find AE, so first we need <*C*: $_m$<*C* = $\tan^{-1}(\frac{AB}{3AB}) = \tan^{-1}(\frac{1}{3})$ = 18.43494882°. Then $_m$<*B* = 90 – 18.43494882° = 71.56505118°. Now we let $_m$<DAE = *x*°. This makes $_m$<EAB = (90 – *x*)°. We also know that EB = CE = 6.454972242 (because AE is a median). We can set up two equations using triangles AEC and ABE (implementing the *Law of* Sines) and then solve for *x*, and this will help us find AE. The first equation will be $\frac{AE}{(71.56505118)^o} = \frac{6.454972242}{(90-x)^o}$, and the second equation is $\frac{AE}{(18.43494882)^o} = \frac{6.454972242}{x^o}$. After isolating AE in both equations, we can set them equal: $\frac{6.454972242(18.43494882)}{x} = \frac{6.454972242(71.56505118)}{90-x}$, which simplifies to 6.454972242(18.43494882)(90 – *x*) = 6.454972242(71.56505118)*x*. Solving for *x*, we get *x* = 18.43494882 degrees. This is also equal to angle *C*, so that AE

(base for triangle ADE) = CE = 6.454972242. Since $_m$<CDE = $\frac{180-18.43494882}{2}$ = 80.78252559 degrees, $_m$<ADE = 180 – 80.78252559 = 99.21747441 degrees. Now we find the height *h* from AE to D using the *Law of Sines* once more (note that the height makes a right triangle that we use now): $\frac{\sin(90^o)}{AD} = \frac{\sin((18.43494882)^o)}{h}$. Since AD = AC – DC = 12.24744871 – 6.454972242 = 5.792476468, we plug this value into the equation, and get *h* = 1.831741893. Area of ΔADE is $\frac{1.831741893(6.454972242)}{2} \approx$ 6.

Answer: 6

219. **According to the figure of the quadrilateral below, what must be true?**

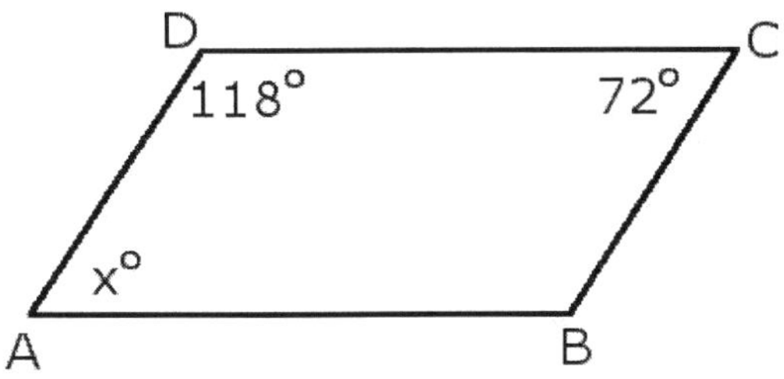

A. AD is not parallel to BC

B. If AB || DC, then $_m$<*x* = 62 degrees

C. the quadrilateral is not a parallelogram

D. all of the above

For AD to be parallel to BC, measures of angles ADC and BCD must add up to 180 degrees. This is not the case in the figure. It can be easily understood if DC is a transversal intersecting AD and BC. Angles ADC and BCD are interior angles in this scenario, which must be equal for AD to be parallel to BC. If AB is parallel to DC, then the angle measure sum of measures of angles *x* and ADC is 180 degrees, which means that $_m$<*x* = 180 – 118 = 62 degrees. Since AD is not parallel to BC, the quadrilateral is not a parallelogram.

Answer: D

220. **In the following figure, ABCD is a square. Segments AE and BE are drawn from the square vertices A and B. If *E* is a midpoint of DC, what is the measure of <AEB (nearest whole)?**

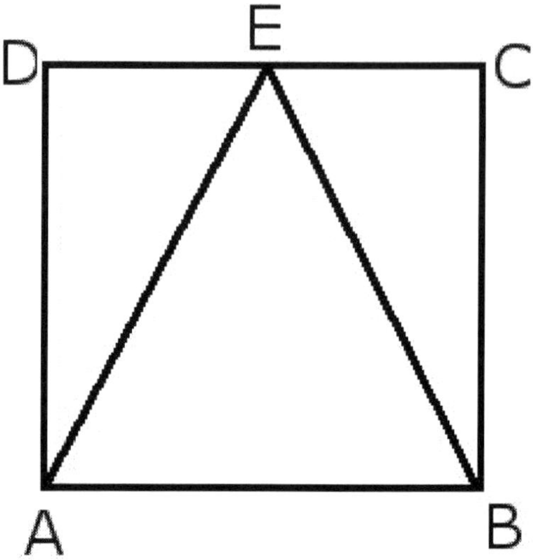

A. 59°

B. 57°

C. 51°

D. 53°

Since segments AE and BE are equal and meet at midpoint E of DC, and AD = BC, angles DEA and CEB must be equal. We then have $_{m}$<AEB = 180 – $_{m}$<DEA – $_{m}$<CEB = 180 – 2$_{m}$<DEA. If we let AD be x, then DE = $\frac{x}{2}$. Using trigonometry, we have $_{m}$<DEA = $\tan^{-1}\left(\frac{x}{\frac{x}{2}}\right) = \tan^{-1}(2)$ = 63.43494882 degrees. This makes angle AEB be 180 – $2(63.43494882^{o}) \approx 53$ degrees.

Answer: D

221. **The following figure shows a circle containing points *A*, *D* and *C*. If the central angle *ABC* of the circle measures 94°, and AD = DC, what is the measure of <*DAB*?**

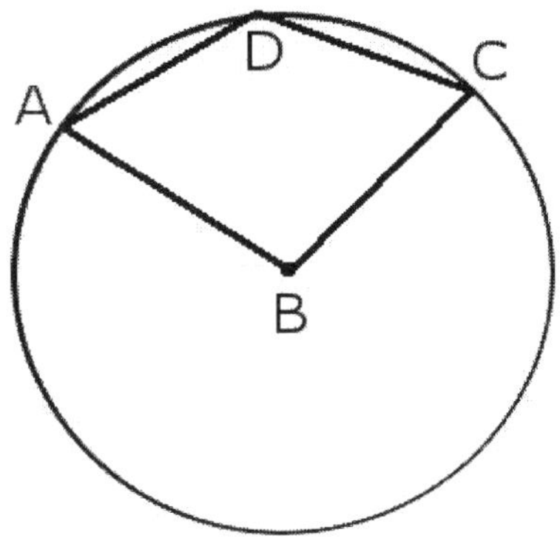

A. 65.5°

B. 66°

C. 66.5°

D. 67°

Since AD = DC, and AB = BC, then <DAB = <BCD. Since $_m$<ABC = 94 degrees, minor arc AC also measures 94 degrees. The major arc AC measures 360 – 94 = 266 degrees, which means that angles ADC measures $\frac{266}{2}$ = 133 degrees. Thus, $_m$<DAB = $\frac{360-94-133}{2}$ = 66.5°.

Answer: C

222. **A page whose length is twice the width has an area of 12 square units. If the shaded region of the page has margins with uniform thickness of 1 unit, what is the area of the shaded region?**

A. $16 + 6\sqrt{6}$

B. $16 - 6\sqrt{6}$

C. $12 - \sqrt{6}$

D. $12 + \sqrt{6}$

Let x be the length of the page, and y be the width. Then $x = 2y$. Since $xy = 12$, then $2y^2 = 12$, so that $y = \sqrt{6}$, and $x = 2\sqrt{6}$. Since the shaded region has uniform thickness margins, the length of the shaded region is $2\sqrt{6} - 2$, and width is $\sqrt{6} - 2$. Thus its area is $(2\sqrt{6} - 2)(\sqrt{6} - 2) = 12 - 6\sqrt{6} + 4 = 16 - 6\sqrt{6}$.

Answer: B

223. **If the volume of the cone is 54, find the base radius of the cone (nearest whole) if its vertical height is 3 times its base radius.**

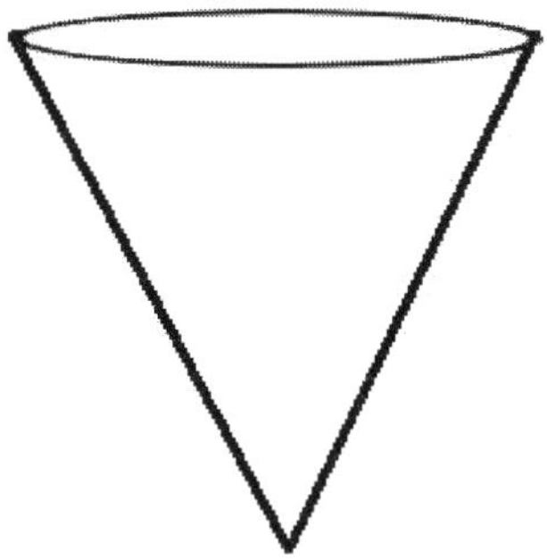

A. 1

B. 2

C. 3

D. 4

If h is the vertical height of the cone, and r is the base radius of the cone, then $h = 3r$.

Since the volume formula is $\frac{\pi r^2 h}{3} = \frac{\pi r^2 (3r)}{3} = \pi r^3$. Since this is 54, $r = \left(\frac{54}{\pi}\right)^{\frac{1}{3}} \approx 3$.

Answer: C

224. **The following figure shows a square containing four identical right triangles. Segment AB is twice the segment CD. Find the area of the square.**

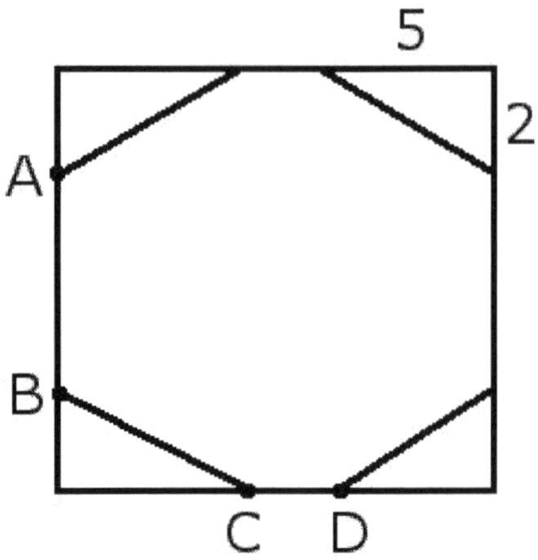

A. 230

B. 252

C. 16

D. 256

If we let CD = x, then AB = $2x$. One side of the square is $2(2) + 2x$, and the side perpendicular to it $2(5) + x$. Setting these two sides equal, we have $4 + 2x = 10 + x$, so that $x = 6$. The side of the square is then $10 + 6 = 16$, and the area of the square is $16^2 = 256$.

Answer: D

225. **A right circular cone has a lateral surface area of 448 square inches. If the radius of the cone is 7 inches, find the vertical height to the nearest whole of an inch.**

A. 8 in

B. 9 in

C. 10 in

D. 11 in

We have 448 = $7\pi\left(7+\sqrt{h^2+49}\right)$, so that 448 = 49π + $7\pi\sqrt{h^2+49}$, and $\sqrt{h^2+49}$ = $\frac{448-49\pi}{7\pi}$. Squaring both sides and simplifying we get $h \approx 11$ inches.

Answer: D

226. **One base of a trapezoid is $\frac{3}{4}$ the other base, and the vertical height is 4. If the area of the trapezoid is 52, find the measure of the dashed line (nearest whole).**

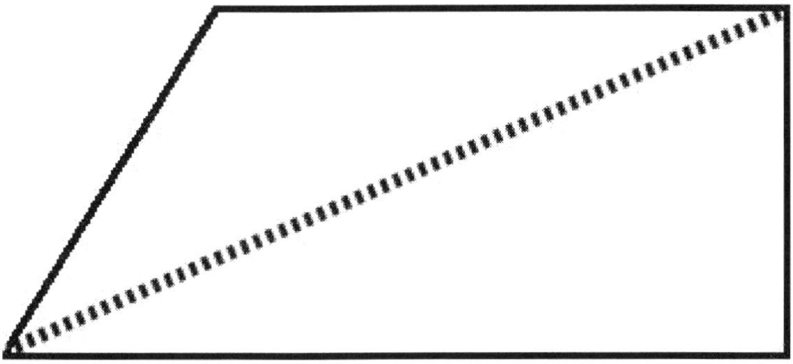

A. 12

B. 13

C. 14

D. 15

Let the bottom base of the trapezoid be *x*. Then the top base is $\frac{3x}{4}$. The sum of the two bases is $\frac{3x}{4} + x = \frac{7x}{4}$. The area of the trapezoid is $52 = \frac{4\left(\frac{7x}{4}\right)}{2}$, giving $x = \frac{104}{7}$. Using the *Pythagorean Theorem* and now known base and height, we have the dashed line as $\sqrt{\left(\frac{104}{7}\right)^2 + 16} \approx 15$.

Answer: D

227. **A floor has a shape of a trapezoid inscribing a square that inscribes a circle that is tangent to the square at its four side midpoints. Trapezoid area is 30, and area of the square is 20. If a ball is randomly dropped on the floor from the ceiling, find the probability that it hits the shaded region of the floor.**

A. $\frac{1}{3} + \frac{\pi}{3}$

B. $\frac{2}{3} - \frac{\pi}{6}$

C. $\frac{2}{3} - \pi$

D. $\frac{1}{3} + \frac{\pi}{6}$

Each side of the square is $\sqrt{20} = 2\sqrt{5}$, which means that the radius of the circle is $\sqrt{5}$. Area of the circle is $\pi(\sqrt{5})^2 = 5\pi$. The combined area of the four shaded regions is $20 - 5\pi$. Since the area of the trapezoid is 30, the probability that the ball lands on any one of the four shaded regions is $\frac{20 - 5\pi}{30} = \frac{2}{3} - \frac{\pi}{6}$.

Answer: B

228. **Richard and Brad start walking in opposite directions in a straight line path from the same point to their homes. If Richard walks north 2 miles and then west 3 miles, and Brad walks south 4 miles and then east 5 miles, find the straight line distance, in miles to the nearest whole, between their homes.**

A. 10

B. 9

C. 8

D. 7

We use the *Pythagorean Theorem* for this problem. Richard's home from the starting point is $\sqrt{2^2 + 3^2} = \sqrt{13}$ miles. Brad's home from the starting point is $\sqrt{4^2 + 5^2} = \sqrt{41}$ miles. Thus, the total distance between their homes is $\sqrt{13} + \sqrt{41} \approx 10$ miles.

Answer: A

229. **An isosceles triangle is inscribed in a circle whose area is 25π square units. If its hypotenuse is equal to the diameter of the circle, find the area of the triangle.**

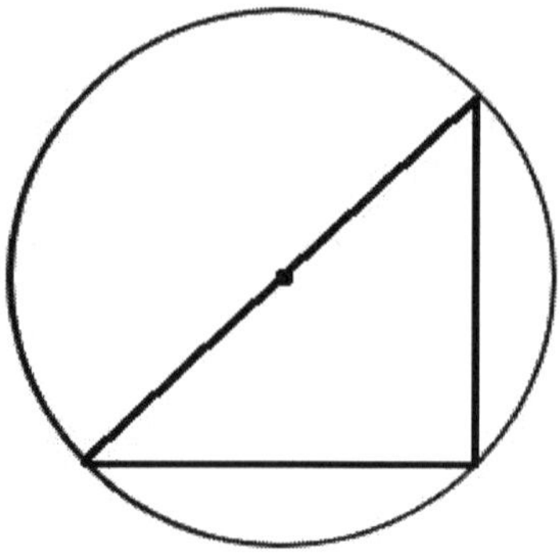

A. 100

B. 5

C. 50

D. 25

The radius of the circle is 5 units (since the area is 25π square units). The diameter is 5(2) = 10 units. The triangle must be a right triangle, because the vertex angle is the inscribed angle that spans a 180-degree arc, thus measuring 90 degrees. We can then use the *Pythagorean Theorem*. Naming each equal side *x*, we have $10 = \sqrt{2x^2}$, so that $x = \frac{10}{\sqrt{2}}$ units. Thus the area of the triangle is $\frac{\left(\frac{10}{\sqrt{2}}\right)^2}{2} = 25$.

Answer: D

230. **A right isosceles triangle is attached to the semi-circle in the figure shown below, making its height equal to the diameter of the semi-circle. If the area of the figure is 64, what is the area of the triangle? Round your answer to the nearest whole.**

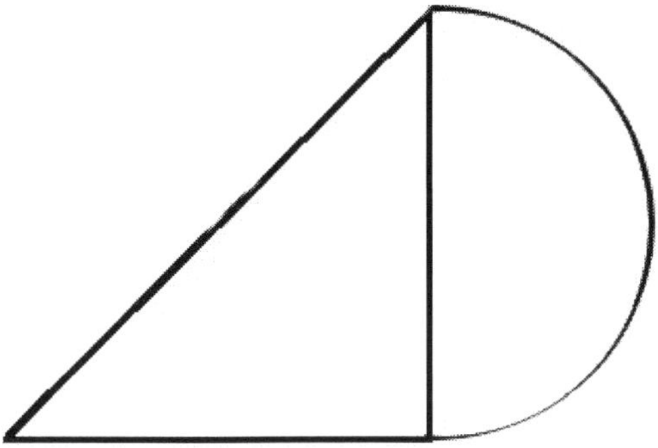

A. 25

B. 35

C. 36

D. 37

Let the side of the triangle be x. Then the triangle area is $\frac{x^2}{2}$, and semi-circle area $\frac{\pi\left(\frac{x}{2}\right)^2}{2} = \frac{\pi x^2}{8}$. Now, we have $64 = \frac{x^2}{2} + \frac{\pi x^2}{8} = x^2(\frac{1}{2} + \frac{\pi}{8})$, making $x^2 = \frac{64}{\frac{1}{2}+\frac{\pi}{8}} =$ 71.69269165. This makes the triangle area $\frac{71.69269165}{2} \approx 36$.

Answer: C

231. **The height of a right circular cylinder is four times its diameter. If the volume of the cylinder is 64π, find the circumference of one of the bases of the cylinder.**

A. 1

B. π

C. 4

D. 4π

Let x be the diameter of the cylinder. The radius is $\frac{x}{2}$ and height h is $4x$. Thus $64\pi = \pi(\frac{x}{2})^2(4x)$, so that $x = 4$. This is the diameter of the cylinder. Thus the circumference of the cylinder base is simply 4π.

Answer: D

232. **The following figure consists of 2 fourths of the circumferences of two identical and tangent circles and a line segment that is 4 units long. Find the area of the figure.**

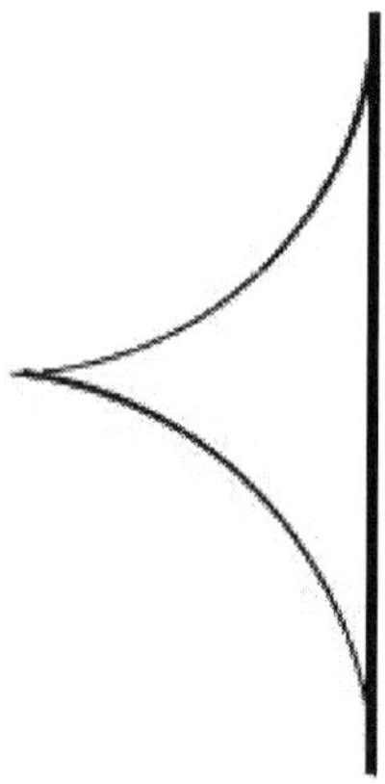

A. $8 - \pi$

B. $8 - 2\pi$

C. $4 - 2\pi$

D. $4 - \pi$

The area of the figure can be found by subtracting the area taken by the two quarter-circles from the area of the rectangle. See the following figure for clarification.

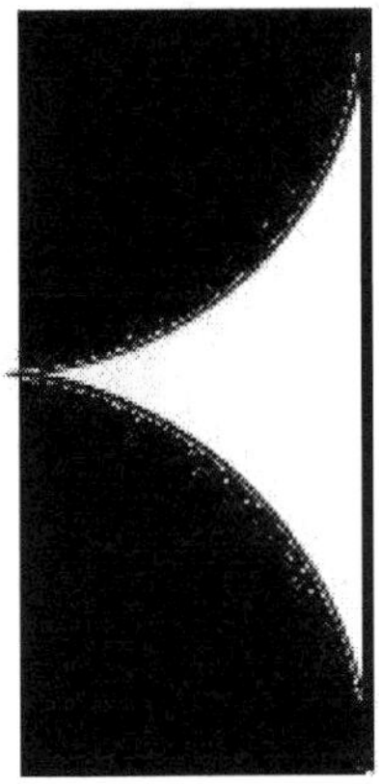

The shaded area is the area defined by the two quarter-circles, and this area must be subtracted from the entire rectangle to find the area of the original figure. Since the line segment covers the distance of the circle diameter, the radius of the circle is 2 units. This is the width of the rectangle. The area of the rectangle is 4 • 2 = 8, and the two areas defined by the two quarter-circles is the area of one semi-circle with the same radius. This area is $\frac{\pi(2^2)}{2}$ = 2π. Thus the area of the original figure is 8 – 2π.

Answer: B

233. **Equation of a line that makes an area of 25 with line equations *y* = -*x* + 5 and *y* = 0 could be**

A. *x* = -5

B. *x* = 0

C. *y* = -*x* – 5

D. *y* = *x* + 5

We have to find an equation so that it serves as a bounded region along with the two given line equations. The first choice does not work, because *x* = -5 line makes an area $10 \cdot \frac{10}{2}$ = 50. We see that through noting the base of the triangle (absolute value of 5 to the left of the origin, on the x-axis, and 5 to the right, determining the base of the triangle formed with equation $y = -x + 5$, which is 10), and height determined by where the lines $x = -5$ and $y = -x + 5$ meet (at $y = 10$). Choice B does not work, because $x = 0$ makes a triangle whose base and height are 5, this producing area 12.5. Choice C does not work, because lines $y = -x + 5$ and $y = -x - 5$ never meet (they are parallel). Choice D works, because the x-axis (line $y = 0$) makes a base 10 since the line $y = x + 5$ touches the x-axis at $x = -5$, and line $y = -x + 5$ touches the x-axis at $x = 5$. Since the height is 5 (the equations $y = x + 5$ and $y = -x + 5$ meet at $y = 5$), the area of the triangle is $10 \cdot \frac{5}{2} = 25$.

Answer: D

234. **A rectangle with perimeter 40 is one-third as wide as it is long. What is the measure of its diagonal (nearest whole)?**

A. 13

B. 14

C. 15

D. 16

If the length of the rectangle is x, then the width is $\frac{x}{3}$. Thus $40 = 2x + \frac{2x}{3}$, making $x = 15$ (length) and $\frac{15}{3} = 5$ (width). Using the Pythagorean Theorem, the diagonal is calculated to be $\sqrt{15^2 + 5^2} \approx 16$.

Answer: D

235. **Find the coordinates of the center of the rectangle if its bottom length is determined by its endpoints (-1, 2) and (5, 2), and its width measures 5 units.**

A. $(5, \frac{9}{2})$

B. $(\frac{9}{2}, 2)$

C. $(2, \frac{9}{2})$

D. $(\frac{9}{2}, 5)$

Since the bottom length is given, the width must be drawn up, thus the possible coordinate of the second endpoint for the width is (5, 2 + 5) = (5, 7). The center of the rectangle is a midpoint between the endpoints of its diagonal. These endpoints are (5, 7) (what we just found), and (-1, 2). Thus the center point of the rectangle is $(\frac{-1+5}{2}, \frac{2+7}{2}) = (2, \frac{9}{2})$.

Answer: C

236. **In a line connecting points *A*, *B*, *C*, and *D*, *B* is between *A* and *C*, and *C* is between *B* and *D*. Segment *AB* measures 5 units. If segment *AC* is three times the length of *AB*, and *CD* is one-half of *BC*, what is the measure of the segment formed by the midpoints of *AB* and *CD*?**

A. 14.5

B. 15

C. 10

D. 15.5

Since $AC = 3AB = 3(5) = 15$, then $BC = AC - AB = 15 - 5 = 10$. This makes $CD = \frac{10}{2} = 5$. Since the segment we need to find consists of one-half of *AB*, one-half of *CD* and full *BC*, we have it as $10 + \frac{5}{2} + \frac{5}{2} = 10 + \frac{10}{2} = 10 + 5 = 15$.

Answer: B

237. **In the following figure, if AB || CD, find $_m$<BDC.**

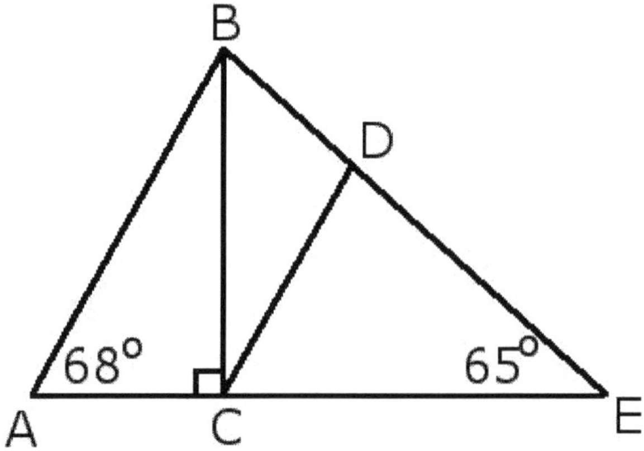

A. 139°

B. 136°

C. 130°

D. 133°

Since ΔABC is a right triangle, $_m$<ABC = 90 – 68 = 22°. Since AB is parallel to CD, we can think of BC as a transversal, which makes angles ABC and BCD alternate angles, so that $_m$<BCD is also 22 degrees. This makes $_m$<DCE = 180 – 90 – 22 = 68°, and $_m$<BDC = 65 + 68 = 133° (we see this while taking angle BDC as an exterior angle that is a sum of non-adjacent interior angles DCE and E for ΔCDE).

Answer: D

238. **How many complete isosceles triangles, with each having two sides measuring 1.5, can fit inside the rectangle with dimensions 3 and 7?**

A. 16

B. 17

C. 18

D. 19

Since the question requires finding the number of complete triangles, this means that there may or may not be extra space after the complete triangles exhaust the rectangle space. We can simply count the number of squares with side measuring 1.5, each of which has two identical triangles with two sides measuring 1.5. The rectangle length of 7 can fit 4 such squares, and rectangle width of 3 can fit exactly 2 such squares. Thus, there are 2(4) = 8 squares in total that fit into the rectangle, which means that 8(2) = 16 isosceles triangles may fit into the rectangle. This takes 18 units out of 21 units of the total rectangle area. Even though each triangle area is $\frac{(1.5)^2}{2} = 1.125$, and at least two more triangle areas would theoretically fill the remaining space, the space itself does not allow any more than 16 triangles to fit, because we have a 1 by 3 rectangle as the extra space.

Answer: A

239. **The following figure shows a trapezoid. Find the measure of the dashed diagonal. Round your answer to the nearest whole.** **Note: figure not drawn to scale**

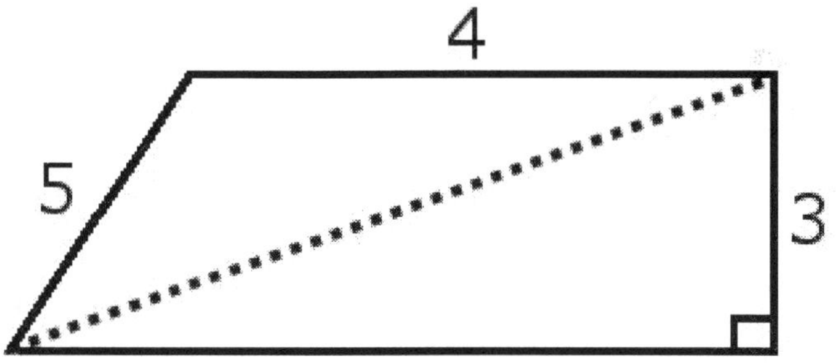

A. 9

B. 8

C. 7

D. 6

We already have the vertical component of the dashed line, which is 3. We need the horizontal component, which consists of the rectangle length measuring 4, and triangle base leg (we split the trapezoid into a rectangle and a triangle). Since the triangle hypotenuse is 5 and height is 3, the triangle is a 3-4-5 triangle, which means that the base leg must be 4 (note that the figure is not drawn to scale). The horizontal component of the dashed line is then 4 + 4 = 8. Thus, using the Pythagorean Theorem, the dashed line is calculated as $\sqrt{8^2 + 3^2} \approx 9$.

Answer: A

240. **A beaker contains 750 mL of water. What must be the width of a rectangular prism-shaped flask that will hold this amount of water if its height is 13 cm and length 12 cm? Round your answer to the nearest tenth of a centimeter.**

A. 4.7

B. 5.0

C. 4.9

D. 4.8

Since 1mL = 1 cm^3, letting the width be w, we have $12 \cdot 13 \cdot w = 750$, which gives $w \approx$ 4.8 cm.

Answer: D

241. **In the following figure, if csc*A* = 3, what is the measure of angle *C*? Round your answer to the nearest whole.**

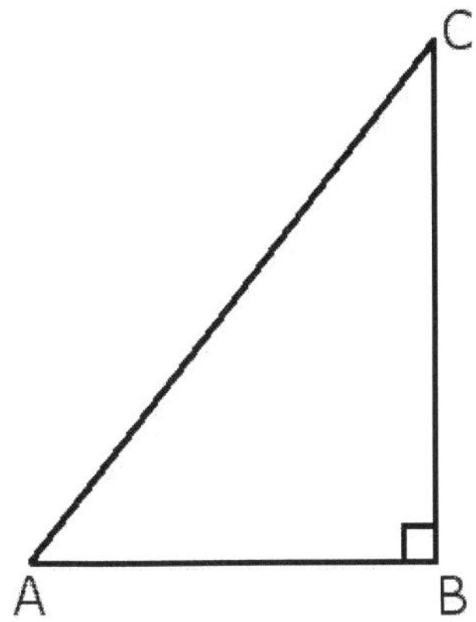

A. 69°

B. 18°

C. 19°

D. 71°

Since csc$A$ = 3, this means that the hypotenuse AC is 3 and BC = 1. Using arccosine, we have $m\angle C = \cos^{-1}(\frac{1}{3}) \approx 71°$.

Answer: D

242. **A triangle has a side ratio 4:11:14. What is the sum of the least and greatest interior angles of this triangle? Round your answer to the nearest whole of a degree.**

A. 93°

B. 68°

C. 112°

D. 25°

In this problem we have to keep in mind that the side ratio also applies to the interior angle ratio. Thus, $4x + 11x + 14x = 180$, making $x = \frac{180}{29}$. This means that the smallest angle measures $4 \cdot \frac{180}{29}$ degrees, and the greatest angle measures $14 \cdot \frac{180}{29}$ degrees. Hence their sum is $4 \cdot \frac{180}{29} + 14 \cdot \frac{180}{29} \approx 112$ degrees.

Answer: C

243. **The volume of a cube is 64. What is the surface area of a similar cube whose side is half the side of the larger cube?**

A. 16

B. 24

C. 36

D. 12

The side of the larger cube is $64^{\frac{1}{3}} = 4$, so that the side of the similar smaller cube is 2. The surface area of the smaller cube is then $6 \cdot 2 \cdot 2 = 24$.

We can also find this result using the fact that the area ratio of similar figures is the square of the side ratio of similar figures. Since the side ratio for similar cubes is 2:4, then the surface area ratio is 4:16 (square of the side ratio). Thus the surface area of the smaller cube is one-fourth of the surface area of the larger cube. Since the surface area of the larger cube is $6 \cdot 4 \cdot 4 = 96$, the surface area of the smaller cube is $\frac{96}{4} = 24$.

Answer: B

244. **If the perimeter of a rhombus is 12 and sum of one pair of its opposite interior angles is 120, what is its area?**

A. $\frac{3\sqrt{3}}{2}$

B. $3\sqrt{3}$

C. $9\sqrt{3}$

D. $\frac{9\sqrt{3}}{2}$

Since the rhombus has equal sides, each side measures $\frac{12}{4}$ = 3 units. Since the sum of one pair of its opposite interior angles is 120, each angle in this pair must be 60 degrees (opposite angles are equal in any rhombus). We now have a triangle we can work with (see the figure below).

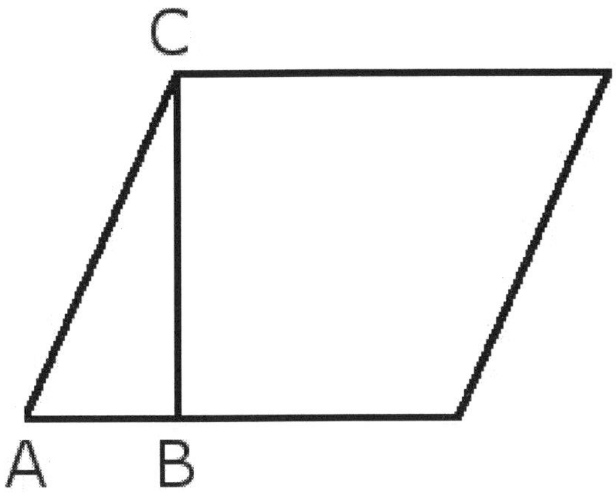

Since AC = 3, and $_m\angle A$ = 60, we can find height BC: sin(60°) = $\frac{BC}{3}$, so that BC = $\frac{3\sqrt{3}}{2}$.

Thus, the area of the rhombus is (3)($\frac{3\sqrt{3}}{2}$) = $\frac{9\sqrt{3}}{2}$.

Another way to find the area of the rhombus is to multiply its diagonals and then divide their product by 2. Since the diagonals are perpendicular to each in any rhombus, and all interior angles are bisected by the diagonals in any rhombus, we can again work with any right triangle created (there are four identical triangles created) by the two diagonals. We can use a triangle with vertices A, C and center of the rhombus as an example triangle. This is a 30-60-90 triangle (angle A is bisected, so 60 becomes 30 degrees). Angle *C* is then 60 degrees in this triangle (makes sense, because original obtuse 120-degree angle *C* is bisected by a shorter diagonal). We can use either the cosine or sine properties to find base and height for this triangle (these are diagonal halves). When we calculate the triangle halves, we can find the area using the formula underlined above. This is left for the reader to execute.

Answer: D

245. **A rectangular 100 by 230 field is changed so that the length is increased by 4 and width is decreased by 4. What is the percent increase/decrease of the new area of the field?**

A. decrease by 2.1%

B. decrease by 2.3%

C. decrease by 2.4%

D. increase by 2.2%

The original area is 230 • 100 = 23,000 square units. The altered field has an area measuring 234 • 96 = 22,464 square units. Therefore the area decreased. The percent of this decrease is $\frac{23{,}000 - 22{,}464}{23{,}000} \cdot 100 \approx 2.3\%$.

Answer: B

246. **In the following figure, if AC = BC and $_m$<*C* = 2$_m$<*B*, what is $_m$<*A*?**

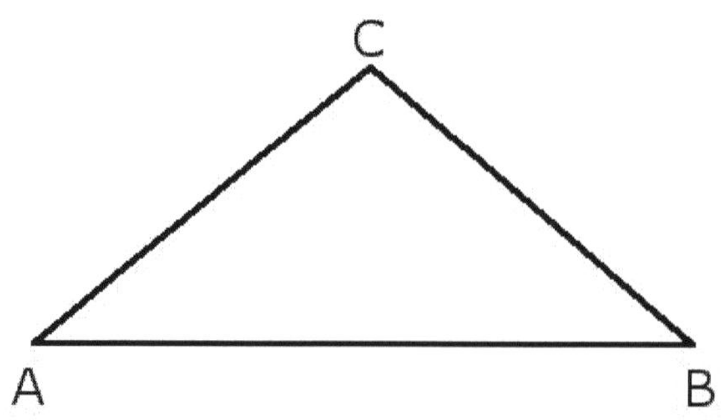

A. 45°

B. 35°

C. 30°

D. 60°

Since AC = BC, $_m\angle A = {}_m\angle B$. Since $_m\angle C = 2_m\angle B$, $2_m\angle B + {}_m\angle B + {}_m\angle B = 180°$, so that $_m\angle B = {}_m\angle A = 45$ degrees.

Answer: A

247. **What is the area of a square whose diagonal endpoints are given by coordinates (-2, 4) and (1, -3)?**

A. $2\sqrt{29}$

B. $\sqrt{58}$

C. $\sqrt{29}$

D. 29

First we find the distance from one endpoint to another (length of the diagonal of the square). This is $\sqrt{(-2-1)^2 + \left(4-(-3)\right)^2} = \sqrt{58}$. Now, let the square side be x. Then, $2x^2 = 58$, so that $x = \sqrt{29}$. Thus the area of the square is $(\sqrt{29})^2 = 29$.

Answer: D

248. **If the area of a square is 25, what is the diagonal half equal to?**

A. 5

B. $5\sqrt{2}$

C. $\frac{\sqrt{2}}{2}$

D. $\frac{5\sqrt{2}}{2}$

Let x be the square side. Then the diagonal measures $\sqrt{x^2 + x^2} = \sqrt{2}x$ units. Since 25 $= x^2$, $x = 5$, and this means that the diagonal measures $5\sqrt{2}$ units. Thus its half measures $\frac{5\sqrt{2}}{2}$ units.

Answer: D

249. **A rectangular region, measuring 25 by 13, consists of a rectangular garden and a uniform thickness 4 walkway surrounding it. What is the measure of diagonal of the actual garden without the walkway? Round your answer to the nearest whole.**

A. 23

B. 24

C. 28

D. 20

The dimensions of the actual garden (without the walkway) are 25 – 4 = 21, and 13 – 4 = 9. Thus the diagonal of the actual garden measures $\sqrt{9^2 + 21^2} \approx 23$.

Answer: A

250. **For what value of b will the line, whose equation is $y = (2b - 2)x + 2$, be perpendicular to the line whose equation is $-2y + 5x - 6 = 0$?**

A. 0

B. 1

C. $\frac{4}{5}$

D. $\frac{8}{5}$

Rewriting the second equation in slope-intercept form, we have $y = \frac{5x}{2} - 3$. The slope of this line is $\frac{5}{2}$. Thus the slope of the perpendicular line to it must have the slope $-\frac{2}{5}$ (the negative reciprocal). This means that $-\frac{2}{5} = 2b - 2$. This makes $b = \frac{4}{5}$.

Answer: C

251. **In the following figure, AB = 7, AD = $\frac{AB}{2}$, and BC = DE. Find the positive difference between measures of angles *x* and *y*. Round your answer to the nearest whole of a degree.** **Note: figure not drawn to scale**

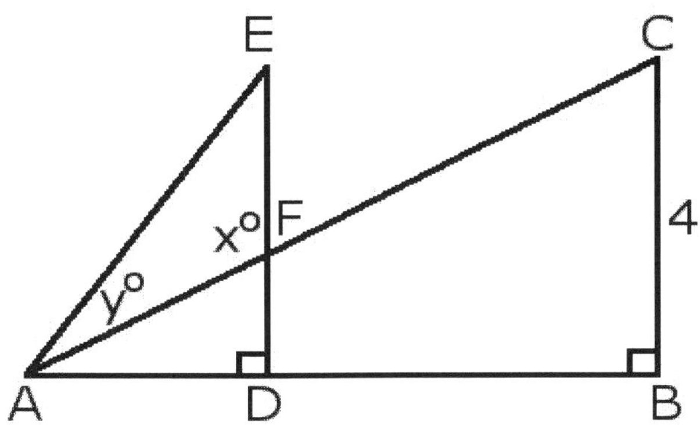

A. 99°

B. 41°

C. 100°

D. 101°

First note that triangles ADF and ABC are similar. Using direct proportion, we have $\frac{BC}{AB}$ = $\frac{FD}{AD}$, and this is $\frac{4}{7}$ = $\frac{FD}{\frac{7}{2}}$, making segment FD = 2. Using trigonometry, we can find the adjacent angle AFD of angle *x*: $_{m}$<AFD = $\tan^{-1}(\frac{AD}{FD})$ = $\tan^{-1}(\frac{\frac{7}{2}}{2})$ = 60.2551187 degrees. This makes $_{m}$<*x* = 180 – 60.2551187 = 119.7448813 degrees. Now, we can find $_{m}$<AED: $\tan^{-1}(\frac{\frac{7}{2}}{4})$ = 41.18592517 degrees, which makes $_{m}$<*y* = 180 – 41.18592517 – 119.7448813 = 19.06919353 degrees. Thus, $_{m}$<*x* – $_{m}$<*y* = 119.7448813 – 19.06919353 ≈ 101°.

Answer: D

252. **A semi-circle is inscribed in a trapezoid. One base of the trapezoid is $\frac{4}{7}$ the other base. If the area of the trapezoid is 100, find the circumference of the semi-circle. Round your answer to the nearest whole.**

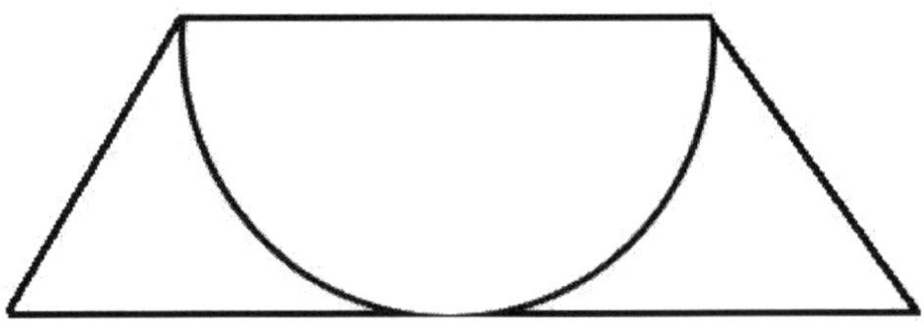

A. 38

B. 18

C. 19

D. 20

Let the semi-circle radius be *r*. Let the longer base of the trapezoid be *b*. Since the diameter of the semi-circle has the same length as the shorter trapezoid base (2*r*), we have $\frac{4b}{7} = 2r$, so that $b = \frac{7r}{2}$. Since the vertical height of the trapezoid has the same length as the radius of the semi-circle (*r*), we have $100 = \frac{r\left(\frac{7r}{2} + 2r\right)}{2}$, which simplifies to $400 = 11r^2$. This makes $r = 6.030226892$. Thus the circumference of the semi-circle is $\pi r = \pi(6.030226892) \approx 19$.

Answer: C

253. **A right shaded circular cone is inscribed in a rhombic prism (see the two-dimensional snapshot below). If the side of the rhombic prism measures 5, find the volume of the cone. Round your answer to the nearest whole.**

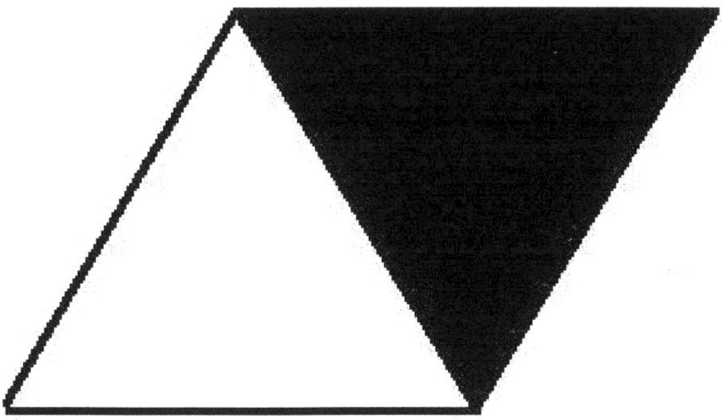

A. 85

B. 9

C. 28

D. 340

Since the cone is a right cone (this is the shaded triangle), its diameter of the base and two slant heights are all equal (all three sides of the shaded triangle must measure 5, including the rhombus' smaller diagonal). We can divide the equilateral shaded triangle in half, creating two right triangles each with hypotenuse 5, base $\frac{5}{2}$ and height (vertical height we need). Using the Pythagorean Theorem, this height must be $\sqrt{5^2 - \left(\frac{5}{2}\right)^2} =$ 4.330127019. The volume of the cone (now that we know its vertical height) is

$$\frac{\pi\left(\left(\frac{5}{2}\right)^2 (4.330127019)\right)}{3} \approx 28.$$

Answer: C

254. **The following figure shows part of the circle whose circumference is 100. If *a* = 2*b*, and there are five (*a* + 3 + *b*) pieces in the entire circle circumference, find the value of (*a* – *b*).**

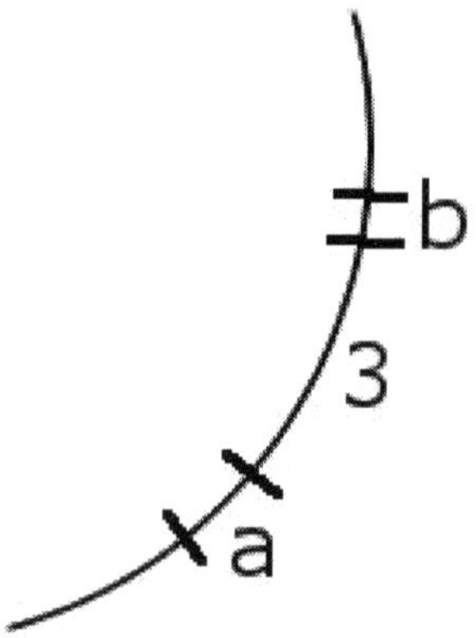

A. 17

B. $\frac{17}{3}$

C. 34

D. $\frac{34}{3}$

Since $a = 2b$, we have $5(2b + 3 + b) = 100$. This makes $b = \frac{17}{3}$ and $a = \frac{34}{3}$. Thus, $a - b = \frac{34}{3} - \frac{17}{3} = \frac{17}{3}$.

Answer: B

255. **Points *A*, *B*, *C*, and *D* are placed in alphabetical order from left to right on the same line. If line segments *AD* = 14, *AB* = 2*CD*, and *AB* = *BC*, what is the ratio of segment *AB* to segment *AD*?**

A. $\frac{1}{14}$:14

B. $\frac{3}{2}$:1

C. $\frac{2}{5}$:1

D. 2:1

Since $AB = 2CD$, then $14 = 2CD + BC + CD$, so that $BC = 14 - 3CD$. Now, $AB = 2CD$ (given). Since $AB = BC$, then $2CD = 14 - 3CD$, which makes $CD = \frac{14}{5}$ and $AB = \frac{28}{5}$. Thus the ratio of AB to AD is $\frac{28}{5}$:14, or $\frac{2}{5}$:1.

Answer: C

256. **A cube-shaped box with volume 64 cubic feet is cut so that each dimension of the new box created is exactly half each dimension of the original box. What is the measure of the longest line segment connecting the vertices of the newly created box?**

A. $\frac{\sqrt{3}}{2}$ feet

B. $2\sqrt{3}$ feet

C. 12 feet

D. $\sqrt{6}$ feet

Each side of the original box measures $64^{\frac{1}{3}}$ = 4 feet. The new box will have a side measuring 2 feet, which makes its longest line segment connecting the vertices $\sqrt{2^2 + 2^2 + 2^2}$ = $\sqrt{12}$ = $2\sqrt{3}$ feet.

Answer: B

257. **The following figure shows a circle with central angle *B*, and points *A* and *C* lying on the circle. If arc *AC* = 70°, find $_{m}$<*BAD*.**

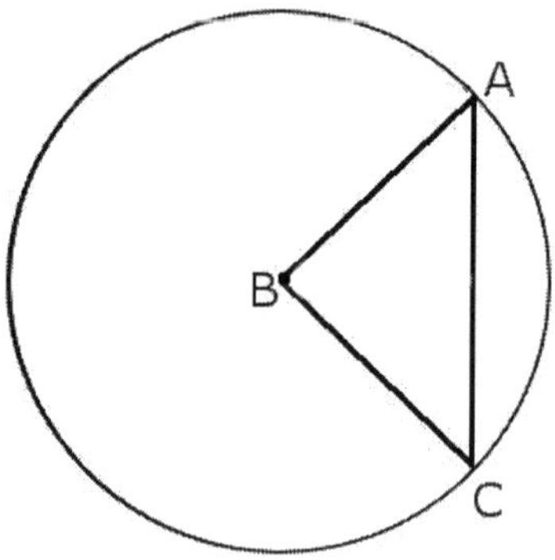

A. 45°

B. 50°

C. 55°

D. 65°

Since <*B* is a central angle, and points *A* and *C* lie on the circle, line segments *AB* and *BC* are circle radiuses (which are of course equal). We conclude that Δ*ABC* is isosceles with congruent angles <*BAD* and <*BCD*. Since the measure of the central angle *B* is the same as the arc that it spans, that is, 70 degrees, m<*BAD* = $\frac{180-70}{2}$ = 55°.

Answer: C

258. **If a cube side measures 4*x* – 3 units, what is its surface area when *x* = 1?**

A. 6

B. 4

C. 2

D. 8

Area of one face of the cube is $(4x-3)^2 = 16x^2 - 24x + 9$. Its surface area is the combined area of all six faces, that is, $6(16x^2 - 24x + 9) = 96x^2 - 144x + 54$ square units. When $x = 1$, this result becomes 96 – 144 + 54 = 6.

Answer: A

259. **The following figure shows two similar triangles. *AB* = 4, *BC* = 6, *AC* = 8, *DF* = 22. What is the perimeter of △DEF?**

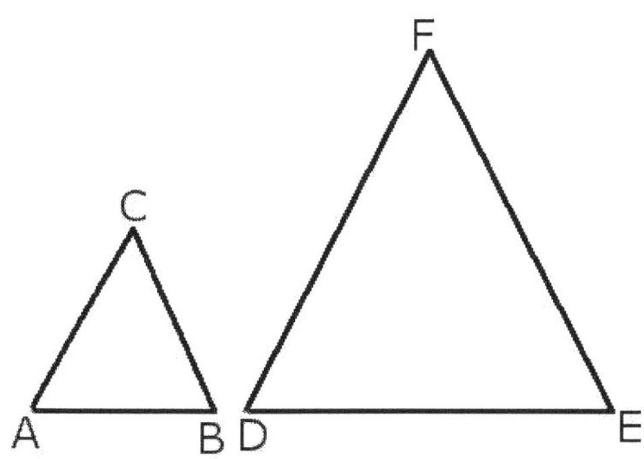

A. $\frac{99}{2}$

B. $\frac{33}{2}$

C. 33

D. 99

The ratio of sides for ΔABC is 4:6:8, or $2x$:$3x$:$4x$ for $x \geq 1$. Side *DF* (22 units in length) corresponds to side *AC* that is 8 units in length (or $4x$ if $x = 2$). Since the triangles are similar, the same ratio of sides applies to ΔDEF. Thus we have $4x = 22$, making $x = \frac{11}{2}$. This makes the sides of ΔDEF measure $2(\frac{11}{2})$, $3(\frac{11}{2})$, and 22 (given), or simply 11, $\frac{33}{2}$ and 22. The perimeter of ΔDEF is therefore $11 + 22 + \frac{33}{2} = 33 + \frac{33}{2} = \frac{99}{2}$.

Answer: A

260. **In the following figure, *AB* = *AC*, and arc *BC* makes a semi-circle. If a line segment *BC* is $\frac{5}{3}$ the line segment *AC*, and area of the semi-circle is 24π, find the perimeter of the figure. Round your answer to the nearest whole.**

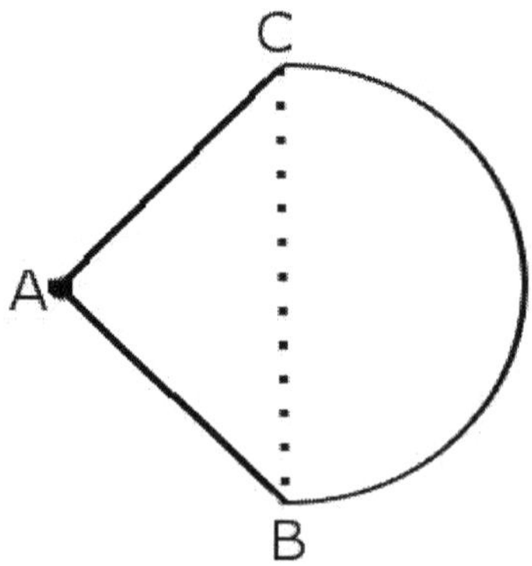

A. 60

B. 27

C. 38

D. 19

Since $AB = AC$, triangle ABC is isosceles. Moreover, we have $BC = \frac{5AC}{3}$. Note that BC is a diameter of the semi-circle. Thus $\frac{BC}{2}$ is the radius, so that the radius in terms of AC is $\frac{5AC}{6}$. Thus we have $24\pi = \frac{\pi\left(\frac{5AC}{6}\right)^2}{2}$, making $AC = \sqrt{69.12}$ and $BC = \frac{5\sqrt{69.12}}{3}$ (this is the diameter of the semi-circle). The circumference of the semi-circle is $\frac{\frac{5\pi\sqrt{69.12}}{3}}{2} = \frac{5\pi\sqrt{69.12}}{6}$. Thus the perimeter of the entire figure is $2\sqrt{69.12} + \frac{5\pi\sqrt{69.12}}{6} \approx 38$.

Answer: C

261. **A slant circular cone is drawn in the figure. If the base radius *AB* = 4, and line segments *AC* = 8 and *BC* = 6, find the volume of the cone. Round your answer to the nearest whole.**

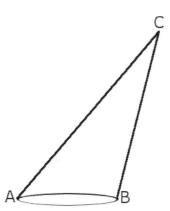

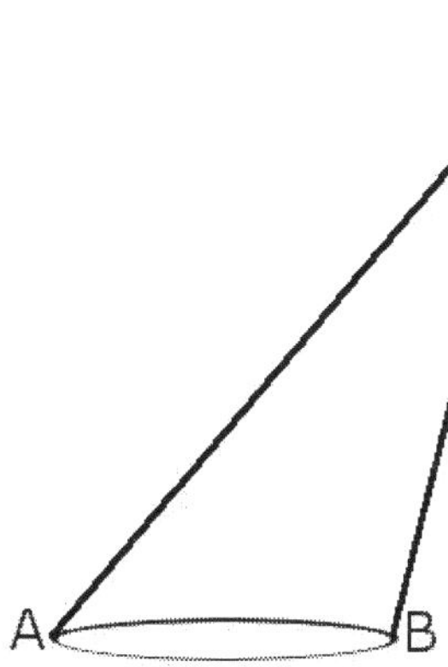

A. 402

B. 134

C. 603

D. 97

If we find the vertical height from vertex *A* to a point on the same horizontal line with vertex *C*, then we can easily find the volume of the cone, because it is simply $\frac{\pi r^2 h}{3}$, where *h* is the vertical height and *r* is the base radius (given). We can extend the two-dimensional shot of a cone (an oblique triangle) to make a right triangle as in the figure below:

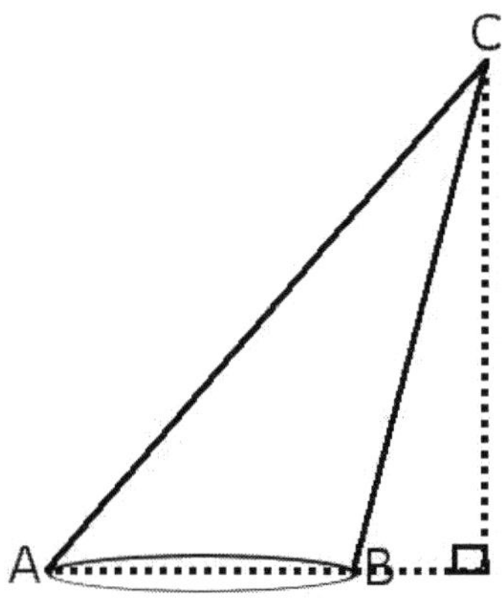

Since we know the measure of *BC*, we can find the vertical height *h* (dashed vertical line segment from vertex *C*) using the measure of the adjacent angle to angle *ABC* and segment *BC*. To find this adjacent angle, we first find $m\angle ABC$ using the Law of Cosines: $AC^2 = AB^2 + BC^2 - 2(AB)(BC)(\cos(m\angle ABC))$, and this is $8^2 = 4^2 + 6^2 - 2(4)(6)\cos(m\angle ABC)$, which makes $\cos(m\angle ABC) = -\frac{1}{4}$, so that $m\angle ABC = \cos^{-1}(-\frac{1}{4}) =$ 104.4775122 degrees. This means that the adjacent angle to this angle is 180 – 104.4775122 = 75.52248781 degrees. Now we can find the vertical height *h* (dashed vertical line from vertex *C*) because we also know *BC*: $\sin(75.52248781^\circ) = \frac{h}{6}$, making $h = 5.809475019$. Thus the volume of the cone is $\frac{\pi(4^2)(5.809475019)}{3} \approx 97$.

Answer: D

262. **The following figure consists of two pairs of identical semi-circles, with the smaller semi-circle diameter equal to the radius of the larger semi-circle. If the vertical line segment *AB* is 5, find the area of the figure.**

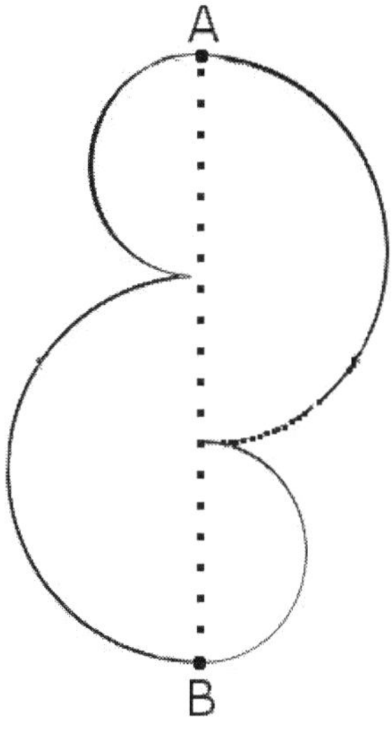

A. $\frac{25\pi}{6}$

B. $\frac{125\pi}{36}$

C. $\frac{125\pi}{18}$

D. $\frac{25\pi}{36}$

Since the line segment *AB* consists of 3 smaller semi-circle diameters, each diameter (and larger semi-circle radius) measures $\frac{5}{3}$ units, and the radius of the smaller semi-circle must be $\frac{5}{6}$ units. Also, the area of the figure is simply the combined area of one smaller complete circle and one larger complete circle. This is $\pi[(\frac{5}{3})^2 + (\frac{5}{6})^2] = \frac{125\pi}{36}$.

Answer: B

263. **A square (shown as a shaded region in the figure below) is taking up a region of a larger square *ABCD* with side 7. If segment *AW* is three times the segment *AX*, find the measure of line segment *XW*. Round your answer to the nearest whole.**

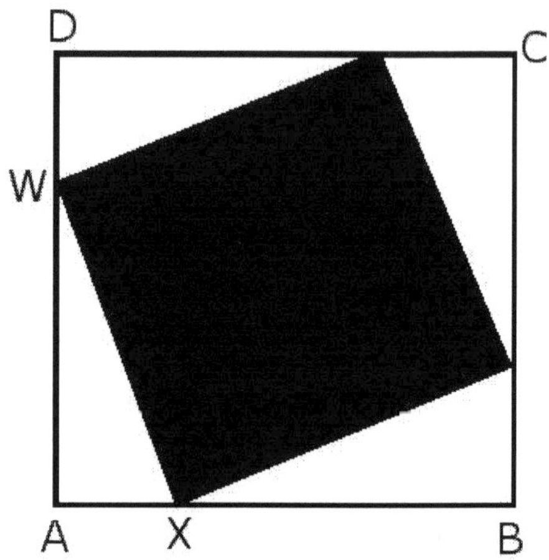

A. 9

B. 8

C. 7

D. 6

Since the shaded square creates four identical right triangles (four non-shaded regions of the larger square), segment *AX* must be equal to segment *DW*. Let the segment *AX* be *x*. Then $3x + x = 7$, making $x = \frac{7}{4} = AX$, and $AW = 3x = \frac{21}{4}$. Since ΔAWX is a right triangle, we can the Pythagorean theorem to the hypotenuse *WX*: $\sqrt{\left(\frac{7}{4}\right)^2 + \left(\frac{21}{4}\right)^2} \approx 6$.

Answer: D

264. **Area of the trapezoid is found by the formula $A = \frac{h(b_1 + b_2)}{2}$. If b_1 is four times the height h, what expression shows b_2 in terms of the area and the height?**

A. $\frac{2A - 4h^2}{h}$

B. $\frac{2A - h^2}{h}$

C. $\frac{A - 4h}{h}$

D. $2A + 4h^2$

We have $b_1 = 4h$. Plugging this into the area formula, we get $A = \frac{h(4h + b_2)}{2}$, or $2A = 4h^2 + hb_2$, which makes $b_2 = \frac{2A - 4h^2}{h}$.

Answer: A

265. **A right circular cylinder has a 5-meter diameter and an 11-meter height. If its density is 140 pounds per cubic meter, what is its mass, in pounds? Round your answer to the nearest whole of a pound.**

A. 38,500 lb

B. 6 lb

C. 30,238 lb

D. 120,951 lb

We first need the volume of the cylinder, which is $\pi(\frac{5}{2})^2(11) = \frac{275\pi}{4}$ cubic meters. Since density = mass/volume, we have the mass as the product of density and volume: 140 lb/m^3 • $(\frac{275\pi}{4})$m^3 ≈ 30,238 lb.

Answer: C

266. **A regular hexagon is divided into six identical triangles, thus creating six central angles. What is the measure of one such central angle?**

A. 55°

B. 40°

C. 45°

D. 60°

A regular hexagon has all six equal sides. Since the six triangles created are identical, the six central angles are also equal. Thus each central angle measures $\frac{360}{6} = 60$ degrees.

Answer: D

267. **In the following figure, 3 identical semi-circles (shown shaded) span each side of the equilateral triangle (diameter of the semi-circles is equal to the equilateral triangle side). If height of the triangle is 4, find the area of the figure. Round your answer to the nearest whole.**

A. 210

B. 9

C. 25

D. 34

Let the equilateral triangle side be x. We can split the triangle in half (equilateral triangle is at least isosceles), creating two identical right triangles, each with height 4 (the two right triangles share this height). The base of one of these triangles is $\frac{x}{2}$, and hypotenuse is x (this is the equilateral triangle side and also the diameter of each of the semi-circles). We can use the Pythagorean Theorem to solve for x: $x^2 = \frac{x^2}{4} + 16$, leading to $x^2 = \frac{64}{3}$ and $x = \sqrt{\frac{64}{3}}$. The radius of each semi-circle is then $\frac{\sqrt{\frac{64}{3}}}{2}$. Area of the figure is then the combined area of the equilateral triangle and three semi-circles:

$$\left(\frac{3}{2}\right)\pi\left(\frac{\sqrt{\frac{64}{3}}}{2}\right)^2 + \sqrt{\frac{64}{3}}\left(\frac{4}{2}\right) \approx 34.$$

Answer: D

268. **A right circular cone is perfectly inscribed in a right circular cylinder (see figure). What is the base radius of the cone if the volume of the cylinder is 380 and its vertical height is 3 times its diameter? Round your answer to the nearest whole.**

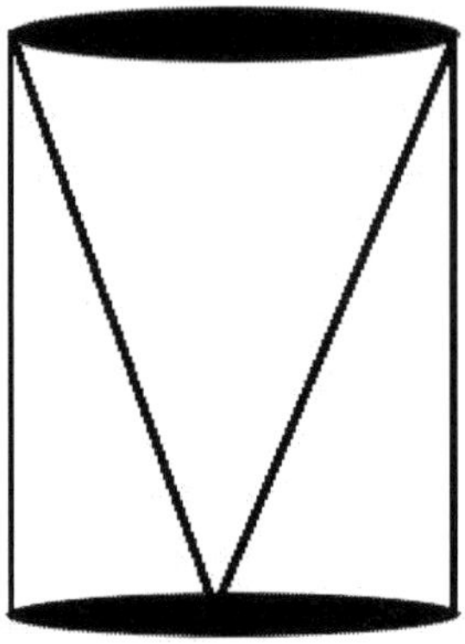

A. 5

B. 4

C. 3

D. 2

First, note that the base radius of the cone is half the cylinder diameter. Let this diameter be x. Then the vertical height of the cylinder, $h = 3x$, and the radius of the cylinder (also the base radius of the cone that we need) is $\frac{x}{2}$. This means that 380 = $\pi(\frac{x}{2})^2(3x)$, so that $\frac{3\pi x^3}{4} = 380$, giving $x = \left(\frac{380(4)}{3\pi}\right)^{\frac{1}{3}}$. Thus $\frac{x}{2} = \frac{\left(\frac{380(4)}{3\pi}\right)^{\frac{1}{3}}}{2} \approx 3$.

Answer: C

77343855R00171

Made in the USA
San Bernardino, CA
22 May 2018